PERIODIC TABLE
OF THE **ELEMENTS**

Transition Metals

PERIODIC TABLE
OF THE **ELEMENTS**

Transition
Metals

**Monica Halka, Ph.D., and
Brian Nordstrom, Ed.D.**

Facts On File
An imprint of Infobase Publishing

TRANSITION METALS

Facts On File, Inc.
An imprint of Infobase Publishing
132 West 31st Street
New York NY 10001

Library of Congress Cataloging-in-Publication Data
Halka, Monica.
   Transition metals / Monica Halka and Brian Nordstrom.
     p. cm. — (Periodic table of the elements)
   Includes bibliographical references and index.
   ISBN 978-0-8160-7371-9
   1. Transition metals. 2. Periodic law. I. Nordstrom, Brian II. Title.
   QD172.T6H35 2011
   546'.6—dc22   2009054139

Facts On File books are available at special discounts when purchased in bulk quantities for businesses, associations, institutions, or sales promotions. Please call our Special Sales Department in New York at (212) 967-8800 or (800) 322-8755.

You can find Facts On File on the World Wide Web at http://www.factsonfile.com

Excerpts included herewith have been reprinted by permission of the copyright holders; the author has made every effort to contact copyright holders. The publishers will be glad to rectify, in future editions, any errors or omissions brought to their notice.

Text design by Erik Lindstrom
Composition by Hermitage Publishing Services
Illustrations by Dale Williams
Photo research by Tobi Zausner, Ph.D.
Cover printed by Sheridan Books, Inc.
Book printed and bound by Sheridan Books, Inc.
Date printed: September 2010
Printed in the United States of America

10 9 8 7 6 5 4 3 2 1

This book is printed on acid-free paper.

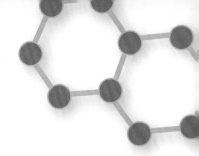

Contents

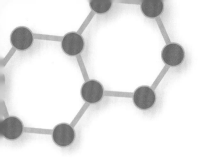

Preface

Speculations about the nature of matter date back to ancient Greek philosophers like Thales, who lived in the sixth century B.C.E., and Democritus, who lived in the fifth century B.C.E., and to whom we credit the first theory of *atoms*. It has taken two and a half millennia for natural philosophers and, more recently, for chemists and physicists to arrive at a modern understanding of the nature of *elements* and *compounds*. By the 19th century, chemists such as John Dalton of England had learned to define elements as pure substances that contain only one kind of atom. It took scientists like the British physicists Joseph John Thomson and Ernest Rutherford in the early years of the 20th century, however, to demonstrate what atoms are—entities composed of even smaller and more elementary particles called *protons, neutrons,* and *electrons.* These particles give atoms their properties and, in turn, give elements their physical and chemical properties.

After Dalton, there were several attempts throughout Western Europe to organize the known elements into a conceptual framework that would account for the similar properties that related groups of elements exhibit and for trends in properties that correlate with increases in atomic weights. The most successful *periodic table* of the elements was designed in 1869 by a Russian chemist, Dmitri Mendeleev. Mendeleev's method of organizing the elements into columns grouping elements with similar chemical and physical properties proved to be so practical that his table is still essentially the only one in use today.

The Russian chemist Dmitri Mendeleev created the periodic table of the elements in the late 1800s. *(HIP/Art Resource)*

While there are many excellent works written about the periodic table (which are listed in the section on further resources), recent scientific investigation has uncovered much that was previously unknown about nearly every element. The Periodic Table of the Elements, a six-volume set, is intended not only to explain how the elements were discovered and what their most prominent chemical and physical properties are, but also to inform the reader of new discoveries and uses in fields ranging from astrophysics to material science. Students, teachers, and the general public seldom have the opportunity to keep abreast of these new developments, as journal articles for the nonspecialist are hard to find. This work attempts to communicate new scientific findings simply and clearly, in language accessible to readers with little or no formal background in chemistry or physics. It should, however, also appeal to scientists who wish to update their understanding of the natural elements.

Each volume highlights a group of related elements as they appear in the periodic table. For each element, the set provides information regarding:

- the discovery and naming of the element, including its role in history, and some (though not all) of the important scientists involved;
- the basics of the element, including such properties as its atomic number, atomic mass, electronic configuration, melting and boiling temperatures, abundances (when known), and important isotopes;
- the chemistry of the element;
- new developments and dilemmas regarding current understanding; and
- past, present, and possible future uses of the element in science and technology.

Some topics, while important to many elements, do not apply to all. Though nearly all elements are known to have originated in stars or stellar explosions, little information is available for some. Some others that have been synthesized by scientists on Earth have not been observed in stellar spectra. If significant astrophysical nucleosynthesis research exists, it is presented as a separate section. The similar situation applies for geophysical research.

Special topic sections describe applications for two or more closely associated elements. Sidebars mainly refer to new developments of special interest. Further resources for the reader appear at the end of the book, with specific listings pertaining to each chapter, as well as a listing of some more general resources.

Acknowledgments

First and foremost, Monica Halka thanks her parents, who convinced me I was capable of achieving any goal. In graduate school, my thesis advisor, Dr. Howard Bryant, influenced my way of thinking about science more than anyone else. Howard taught me that learning requires having the humility to doubt your understanding and that it's important for a physicist to be able to explain her work to anyone. I have always admired the ability to communicate scientific ideas to nonscientists and wish to express my appreciation for conversations with National Public Radio science correspondent Joe Palca, whose clarity of style I attempt to emulate in this work. I also thank my coworkers at Georgia Tech, Dr. Greg Nobles, and Ms. Nicole Leonard, for their patience and humor as I struggled with deadlines.

In 1967, Brian Nordstrom entered the University of California at Berkeley. Several professors, including John Phillips, George Trilling, Robert Brown, Samuel Markowitz, and A. Starker Leopold, made significant and lasting impressions. I owe an especial debt of gratitude to Harold Johnston, who was my graduate research advisor in the field of atmospheric chemistry. Many of the scientists mentioned in the Periodic Table of the Elements set I have known personally: for example, I studied under Neil Bartlett, Kenneth Street, Jr., and physics Nobel laureate Emilio Segrè. I especially cherish having known chemistry Nobel laureate Glenn Seaborg. I also acknowledge my past and present colleagues at California State University, Northern Arizona University, and Embry-Riddle Aeronautical University, Prescott, Arizona, without whom my career in education would not have been as enjoyable.

Finally, both authors thank Jodie Rhodes and Frank Darmstadt for their encouragement, patience, and understanding.

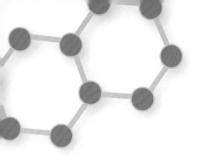

Introduction

This volume is devoted to the chemical and physical properties of an extremely important group of elements—the transition metals—and is intended for readers at the high school or beginning college level who have studied, or who will study, chemistry and physics. Transition metals include some of the world's most industrially significant substances. Iron, copper, silver, gold, and mercury are familiar metals that have played historic roles in civilization for thousands of years. Titanium, chromium, tungsten, and platinum are examples of particularly important elements in today's economy—in some cases adding strength to structural materials such as steel, and in other cases serving as catalysts in chemical reactions. In this book, readers will learn what the important properties of transition metals are and how they are useful in everyday life.

More so than any of the other major groups of elements in the periodic table, the transition metals have shaped human history and been the workhorses of industry. The discovery of metallic copper ended the Stone Age and ushered in the Bronze Age. Alloys of iron (especially steel) later took over, and the Iron Age replaced the Bronze Age. Copper, silver, and gold—and, more recently, platinum—have been the precious metals from which coins and jewelry have been made from ancient times to the present. Mercury was the substance of choice when *incandescent* light bulbs were invented. Thomas Edison chose tungsten to be the filament in light bulbs. Radioactive isotopes of cobalt and technetium continue to be used daily in the diagnosis and treatment of disease in hospital radiology units all around the world. Platinum, palladium, and rhodium are the catalysts in automobile *catalytic converters,* converting toxic exhaust gases into carbon dioxide, nitrogen, and water. Other transition metals that have played vital roles in industry and technology include titanium, vanadium, chromium, manganese, cobalt, nickel, zinc, zirconium, and molybdenum.

What distinguishes a transition metal from other elements in the periodic table is a partially filled *d* subshell of electrons. Since a *d* subshell can hold up to 10 electrons, there are 10 columns of transition metals. Located in the middle of the periodic table, the transition metals are arranged in four rows that correspond to the principal energy levels of the *d* subshells that are being progressively filled with electrons. The elements scandium (number 21) through zinc (number 30) are said to be in the 3d row of the transition metals because electrons are filling the *d* subshell of the third principal energy level. The elements yttrium (number 39) through cadmium (number 48) are in the 4d row. The elements lanthanum (number 57) and hafnium (number 72) through mercury (number 80) are in the 5d row. Finally, synthetic elements actinium (number 89) and rutherfordium (number 104) through copernicium (number 112) are in the 6d row. The gaps in atomic number between lanthanum and hafnium, and again between actinium and rutherfordium, are due to the lanthanide and actinide series of elements. Those elements are separated into their own block of elements (referred to as the *inner transition elements,* or alternatively as the *rare earths* or the *f* block).

The similarities in properties of the transition metals are due to the nature of the valence electrons in the *d* subshells. Transition metals tend to have high melting and boiling points. They are good conductors of heat and electricity, with copper, silver, and gold being exceptionally good examples. Most of the metals in the 3d row tend to dissolve readily in hydrochloric acid (copper being an exception). The heavier metals, however, tend to be relatively *inert* toward acids. Most notable in this regard is the *platinum group* of metals, which includes platinum, palladium, and their neighbors in the periodic table.

An important consequence of having valence *d* electrons is that, as a general rule, transition metals exhibit a much greater number of *oxidation states* than *main group* metals do. The alkali metals only form "+1" ions. The alkaline earths only form "+2" ions. Aluminum and gallium form compounds and ions only in the "+3" oxidation state. Tin and lead form compounds and ions only in the "+2" and "+4" states. Bismuth exhibits only the "+3" and "+5" states. In contrast, it is not uncommon for transition metals to form compounds or ions in which they might exhibit four or five different oxidation states. As a consequence, there is a much greater variety of compounds that can be formed with transition metals, along with more complex chemical bonding than tends to occur elsewhere in the periodic table.

Each chapter discusses a group of elements, including their similarities and differences, current research, and applications.

Chapter 1 covers scandium and yttrium, both important in scientific research. Scandium is also of interest for its use in structural materials, as a strengthening agent and an anticorrosive. Yttrium is used in television picture tubes and flat plasma screens, in cobalt-based alloys, and has applications in optics, acoustics, and radar.

Chapter 2 explores the titanium and vanadium groups, whose elements find important uses in the aerospace industry, aluminum alloys, and particle accelerators. The main use of titanium is to strengthen alloys and to promote corrosion resistance in applications that include airframes, engines, and submarines. Recent research and development has given rise to a bulk metallic glass made from pure zirconium—the first time a glass has been fabricated from a single element. The main use of hafnium is in the control rods of nuclear reactors. Hafnium is also used in gas-filled incandescent light bulbs and in tungsten and molybdenum alloys, and is challenging silicon as the semiconductor of choice in computer chips. More than half the vanadium produced in the United States is alloyed with steel to increase strength and toughness. Niobium is used in stainless steel, high-temperature alloys, and superconducting alloys. Most tantalum is used to make capacitors in applications that include smoke detectors, heart pacemakers, and automobiles.

Chapter 3 investigates the chromium and manganese groups. Chromium has several familiar applications that include chrome plating and the production of stainless steel. There are a number of important alloys that utilize molybdenum and tungsten. Tungsten's best-known application is its use as the filament in incandescent light bulbs. The major use of molybdenum is its addition to steel to increase hardness, toughness, and corrosion resistance. Manganese is an essential component of most forms of steel and may also be used to strengthen brass and to color bricks. Rhenium has numerous uses as a catalyst. Technetium's radioactivity both limits its use and provides its most valuable uses. The main isotope of technetium, Tc-99, has important applications in radiology, particularly in diagnostic tests.

Chapter 4 covers the historically important iron, cobalt, and nickel groups. Iron, mostly as steel, is one of the most important metals used by modern industry and technology. Compounds of iron are added to plant fertilizers and used in nutritional supplements for people with iron deficiencies. Second only to iron, cobalt is an important metal in permanent magnets. In addition, cobalt alloys are used in jet engines in the aerospace industry and in high-speed tools. Nickel is familiar to most people as the nickel coin, or 5-cent piece, currently minted in both the United States and Canada. Nickel has several other uses in the metallic state. It is added to stainless steel to increase corrosion resistance, and copper-nickel alloys are used in water *desalination* plants. A particularly important area of application of the platinum metals is in catalysts. For example, platinum and rhodium are widely used in catalytic converters in motor vehicles. Platinum metal catalysts can also be used to remove trace impurities in other products.

Chapter 5 explores the copper group—copper, silver, and gold. Beyond their uses in coinage and ornamentation, these elements are important in many other everyday activities. Because of its good thermal and electrical conductivity, copper is used extensively in cookware and electrical wiring. Silver or alloys of silver are important components of some batteries, electrical contacts, and printed circuits. Gold is used in computer components. The automobile industry is beginning to use gold in catalytic converters. In addition, the use of gold catalysts in fuel cells is being investigated. Gold bonding wire and gold electroplating are used in electronics and electrical contacts. Research into uses for gold in nanotechnology research is ongoing.

In chapter 6, the reader will find information on the zinc group, which includes zinc, cadmium, and mercury. Zinc is used in batteries, alloys used to make coins, cathodic protection of steel, and is alloyed with copper to make brass. Cadmium is probably best known for its use with nickel in rechargeable Ni-Cad batteries. It can also be alloyed with silver to form a low-melting-point solder. Mercury is probably best known for its use in thermometers: It is also used in dental and medical equipment. Sphygmomanometers are instruments that measure blood pressure and constitute the single largest use of mercury in the medical industry.

Chapter 7 presents possible future developments that involve the transition metals.

Transition Metals provides the reader, whether student or scientist, with an overall up-to-date understanding regarding each of the transition metals—where they came from, how they fit into our current technological society, and where they may lead us.

Overview:
Chemistry and
Physics Background

What *is* an element? To the ancient Greeks, everything on Earth was made from only four elements—earth, air, fire, and water. Celestial bodies—the Sun, moon, planets, and stars—were made of a fifth element: ether. Only gradually did the concept of an element become more specific.

An important observation about nature was that substances can change into other substances. For example, wood burns, producing heat, light, and smoke and leaving ash. Pure metals like gold, copper, silver, iron, and lead can be smelted from their ores. Grape juice can be fermented to make wine and barley fermented to make beer. Food can be cooked; food can also putrefy. The baking of clay converts it into bricks and pottery. These changes are all examples of chemical reactions. Alchemists' careful observations of many chemical reactions greatly helped them to clarify the differences between the most elementary substances ("elements") and combinations of elementary substances ("compounds" or "mixtures").

Elements came to be recognized as simple substances that cannot be decomposed into other even simpler substances by chemical reactions. Some of the elements that had been identified by the Middle Ages are easily recognized in the periodic table because they still have chemical symbols that come from their Latin names. These elements are listed in the table on page xx.

ELEMENTS KNOWN TO ANCIENT PEOPLE

Iron: Fe ("ferrum")	Copper: Cu ("cuprum")
Silver: Ag ("argentum")	Gold: Au ("aurum")
Lead: Pb ("plumbum")	Tin: Sn ("stannum")
Antimony: Sb ("stibium")	Mercury: Hg ("hydrargyrum")
*Sodium: Na ("natrium")	*Potassium: K ("kalium")
Sulfur: S ("sulfur")	

Note: *Sodium and potassium were not isolated as pure elements until the early 1800s, but some of their salts were known to ancient people.

Modern atomic theory began with the work of the English chemist John Dalton in the first decade of the 19th century. As the concept of the atomic composition of matter developed, chemists began to define elements as simple substances that contain only one kind of atom. Because scientists in the 19th century lacked any experimental apparatus capable of probing the structure of atoms, the 19th-century model of the atom was rather simple. Atoms were thought of as small spheres of uniform density; atoms of different elements differed only in their masses. Despite the simplicity of this model of the atom, it was a great step forward in our understanding of the nature of matter. Elements could be defined as simple substances containing only one kind of atom. Compounds are simple substances that contain more than one kind of atom. Because atoms have definite masses, and only whole numbers of atoms can combine to make molecules, the different elements that make up compounds are found in definite proportions by mass. (For example, a molecule of water contains one oxygen atom and two hydrogen atoms, or a mass ratio of oxygen-to-hydrogen of about 8:1.) Since atoms are neither created nor destroyed during ordinary chemical reactions ("ordinary" meaning in contrast to "nuclear" reactions), what happens in chemical reactions is that atoms are rearranged into combinations that differ from the original reactants, but in doing so, the total mass is conserved. Mixtures are combinations of elements that are not in definite proportions. (In salt water, for example, the salt could be 3 percent by mass, or 5 percent by mass, or many other possibilities; regardless of the percentage of salt, it would still be called "salt water.") Chemical reactions are not required to separate the components of mixtures; the components of mixtures can be separated by physical processes such as distillation, evaporation, or precipitation. Examples of elements, compounds, and mixtures are listed in the following table.

The definition of an element became more precise at the dawn of the 20th century with the discovery of the proton. We now know that an atom has a small center called the "nucleus." In the nucleus are one or more protons, positively charged particles, the number of which determine an atom's identity. The number of protons an atom has is referred to as its "atomic number." Hydrogen, the lightest element, has an atomic number of 1, which means each of its atoms contains a single proton. The next element, helium, has an atomic number of 2, which means each of its atoms contain two protons. Lithium has an atomic number of 3, so its atoms have three protons, and so forth, all the way through the periodic table. Atomic nuclei also contain neutrons, but atoms of the same element can have different numbers of neutrons; we call atoms of the same element with different number of neutrons "isotopes."

There are roughly 92 naturally occurring elements—hydrogen through uranium. Of those 92, two elements, technetium (element 43) and promethium (element 61), may once have occurred naturally on Earth, but the atoms that originally occurred on Earth have decayed

EXAMPLES OF ELEMENTS, COMPOUNDS, AND MIXTURES

ELEMENTS	COMPOUNDS	MIXTURES
Hydrogen	Water	Salt water
Oxygen	Carbon dioxide	Air
Carbon	Propane	Natural gas
Sodium	Table salt	Salt and pepper
Iron	Hemoglobin	Blood
Silicon	Silicon dioxide	Sand

away, and those two elements are now produced artificially in nuclear reactors. In fact, technetium is produced in significant quantities because of its daily use by hospitals in nuclear medicine. Some of the other first 92 elements—polonium, astatine, and francium, for example—are so radioactive that they exist in only tiny amounts. All of the elements with atomic numbers greater than 92—the so-called transuranium elements—are all produced artificially in nuclear reactors or particle accelerators. As of the writing of this book, the discoveries of the elements through number 118 (with the exception of number 117) have all been reported. The discoveries of elements with atomic numbers greater than 111 have not yet been confirmed, so those elements have not yet been named.

When the Russian chemist Dmitri Mendeleev (1834–1907) developed his version of the periodic table in 1869, he arranged the elements known at that time in order of *atomic mass* or *atomic weight* so that they fell into columns called *groups* or *families* consisting of elements with similar chemical and physical properties. By doing so, the rows exhibit

Mendeleev's Periodic Table (1871)

Group Period	I	II	III	IV	V	VI	VII	VIII
1	H=1							
2	Li=7	Be=9.4	B=11	C=12	N=14	O=16	F=19	
3	Na=23	Mg=24	Al=27.3	Si=28	P=31	S=32	Cl=35.5	
4	K=39	Ca=40	?=44	Ti=48	V=51	Cr=52	Mn=55	Fe=56, Co=59 Ni=59
5	Cu=63	Zn=65	?=68	?=72	As=75	Se=78	Br=80	
6	Rb=85	Sr=87	?Yt=88	Zr=90	Nb=94	Mo=96	?=100	Ru=104, Rh=104 Pd=106
7	Ag=108	Cd=112	In=113	Sn=118	Sb=122	Te=125	J=127	
8	Cs=133	Ba=137	?Di=138	?Ce=140				
9								
10			?Er=178	?La=180	Ta=182	W=184		Os=195, Ir=197 Pt=198
11	Au=199	Hg=200	Tl=204	Pb=207	Bi=208			
12				Th=231		U=240		

© Infobase Publishing

Dmitri Mendeleev's 1871 periodic table. The elements listed are the ones that were known at that time, arranged in order of increasing relative atomic mass. Mendeleev predicted the existence of elements with masses of 44, 68, and 72. His predictions were later shown to have been correct.

periodic trends in properties going from left to right across the table, hence the reference to rows as *periods* and name "periodic table."

Mendeleev's table was not the first periodic table, nor was Mendeleev the first person to notice *triads* or other groupings of elements with similar properties. What made Mendeleev's table successful and the one we use today are two innovative features. In the 1860s, the concept of *atomic number* had not yet been developed, only the concept of atomic mass. Elements were always listed in order of their atomic masses, beginning with the lightest element, hydrogen, and ending with the heaviest element known at that time, uranium. Gallium and germanium, however, had not yet been discovered. Therefore, if one were listing the known elements in order of atomic mass, arsenic would follow zinc, but that would place arsenic between aluminum and indium. That does not make sense because arsenic's properties are much more like those of phosphorus and antimony, not like those of aluminum and indium.

To place arsenic in its "proper" position, Mendeleev's first innovation was to leave two blank spaces in the table after zinc. He called the first element eka-aluminum and the second element eka-silicon, which he said corresponded to elements that had not yet been discovered but whose properties would resemble the properties of aluminum and silicon, respectively. Not only did Mendeleev predict the elements' existence, he also estimated what their physical and chemical properties should be in analogy to the elements near them. Shortly afterward, these two elements were discovered and their properties were found to be very close to what Mendeleev had predicted. Eka-aluminum was called *gallium* and eka-silicon was called *germanium*. These discoveries validated the predictive power of Mendeleev's arrangement of the elements and demonstrated that Mendeleev's periodic table could be a predictive tool, not just a compendium of information that people already knew.

The second innovation Mendeleev made involved the relative placement of tellurium and iodine. If the elements are listed in strict order of their atomic masses, then iodine should be placed before tellurium, since iodine is lighter. That would place iodine in a group with sulfur and selenium and tellurium in a group with chlorine and bromine, an arrangement that does not work for either iodine or tellurium.

Therefore, Mendeleev rather boldly reversed the order of tellurium and iodine so that tellurium falls below selenium and iodine falls below bromine. More than 40 years later, after Mendeleev's death, the concept of atomic number was introduced, and it was recognized that elements should be listed in order of atomic number, not atomic mass. Mendeleev's ordering was thus vindicated, since tellurium's atomic number is one less than iodine's atomic number. Before he died, Mendeleev was considered for the Nobel Prize, but did not receive sufficient votes to receive the award despite the importance of his insights.

THE PERIODIC TABLE TODAY

All of the elements in the first 12 groups of the periodic table are referred to as *metals*. The first two groups of elements on the left-hand side of the table are the *alkali metals* and the *alkaline earth metals*. All of the alkali metals are extremely similar to each other in their chemical and physical properties, as, in turn, are all of the alkaline earths to each other. The 10 groups of elements in the middle of the periodic table are *transition metals*. The similarities in these groups are not as strong as those in the first two groups, but still satisfy the general trend of similar chemical and physical properties. The transition metals in the last row are not found in nature but have been synthesized artificially. The metals that follow the transition metals are called post-transition metals.

The so-called *rare earth elements*, which are all metals, usually are displayed in a separate block of their own located below the rest of the periodic table. The elements in the first row of rare earths are called *lanthanides* because their properties are extremely similar to the properties of lanthanum. The elements in the second row of rare earths are called *actinides* because their properties are extremely similar to the properties of actinium. The actinides following uranium are called *transuranium elements* and are not found in nature but have been produced artificially.

The far right-hand six groups of the periodic table—the remaining *main group elements*—differ from the first 12 groups in that more than one kind of element is found in them; in this part of the table we find metals, all of the *metalloids* (or *semimetals*), and all of the *nonmetals*. Not counting the artificially synthesized elements in these groups (elements having atomic numbers of 113 and above and that have not yet been named), these six groups contain 7 metals, 8 metalloids, and 16 nonmetals. Except for the last group—the *noble gases*—each individual group has more than just one kind of element. In fact, sometimes nonmetals, metalloids, and metals are all found in the same column, as are the cases with group IVB (C, Si, Ge, Sn, and Pb) and also with group VB (N, P, As, Sb, and Bi). Although similarities in chemical and physical properties are present within a column, the differences are often more striking than the similarities. In some cases, elements in the same column do have very similar chemistry. Triads of such elements include three of the *halogens* in group VIIB—chlorine, bromine, and iodine; and three group VIB elements—sulfur, selenium, and tellurium.

ELEMENTS ARE MADE OF ATOMS

An atom is the fundamental unit of matter. In ordinary chemical reactions, atoms cannot be created or destroyed. Atoms contain smaller *subatomic* particles: protons, neutrons, and electrons. Protons and neutrons are located in the *nucleus,* or center, of the atom and are referred to as *nucleons.* Electrons are located outside the nucleus. Protons and neutrons are comparable in mass and significantly more massive than electrons. Protons carry positive electrical charge. Electrons carry negative charge. Neutrons are electrically neutral.

The identity of an element is determined by the number of protons found in the nucleus of an atom of the element. The number of protons is called an element's atomic number, and is designated by the letter Z. For hydrogen, Z = 1, and for helium, Z = 2. The heaviest naturally occurring element is uranium, with Z = 92. The value of Z is 118 for the heaviest element that has been synthesized artificially.

Atoms of the same element can have varying numbers of neutrons. The number of neutrons is designated by the letter *N*. Atoms of the same element that have different numbers of neutrons are called *isotopes* of that element. The term *isotope* means that the atoms occupy the same place in the periodic table. The sum of an atom's protons and neutrons is called the atom's *mass number.* Mass numbers are dimensionless whole numbers designated by the letter *A* and should not be confused with an atom's *mass,* which is a decimal number expressed in units such as grams. Most elements on Earth have more than one isotope. The average mass number of an element's isotopes is called the element's atomic mass or atomic weight.

The standard notation for designating an atom's atomic and mass numbers is to show the atomic number as a subscript and the mass number as a superscript to the left of the letter representing the element. For example, the two naturally occurring isotopes of hydrogen are written ^1_1H and ^2_1H.

For atoms to be electrically neutral, the number of electrons must equal the number of protons. It is possible, however, for an atom to gain or lose electrons, forming *ions*. Metals tend to lose one or more electrons to form positively charged ions (called *cations*); nonmetals are more likely to gain one or more electrons to form negatively charged ions (called *anions*). Ionic charges are designated with superscripts. For example, a calcium ion is written as Ca^{2+}; a chloride ion is written as Cl^-.

THE PATTERN OF ELECTRONS IN AN ATOM

During the 19th century, when Mendeleev was developing his periodic table, the only property that was known to distinguish an atom of one element from an atom of another element was relative mass. Knowledge of atomic mass, however, did not suggest any relationship between an element's mass and its properties. It took several discoveries—among them that of the electron in 1897 by the British physicist John Joseph ("J. J.") Thomson, *quanta* in 1900 by the German physicist Max Planck, the wave nature of matter in 1923 by the French physicist Louis de Broglie, and the mathematical formulation of the quantum mechanical model of the atom in 1926 by the German physicists Werner Heisenberg and Erwin Schrödinger (all of whom collectively illustrate the international nature of science)—to elucidate the relationship between the structures of atoms and the properties of elements.

The number of protons in the nucleus of an atom defines the identity of that element. Since the number of electrons in a neutral atom is equal to the number of protons, an element's atomic number also reveals how many electrons are in that element's atoms. The electrons occupy regions of space that chemists and physicists call *shells*. The shells are further divided into regions of space called *subshells*. Subshells are related to angular momentum, which designates the shape of the electron orbit space around the nucleus. Shells are numbered 1, 2,

3, 4, and so forth (in theory out to infinity). In addition, shells may be designated by letters: The first shell is the K-shell, the second shell the L-shell, the third the M-shell, and so forth. Subshells have letter designations, s, p, d, and f being the most common. The *n*th shell has *n* possible subshells. Therefore, the first shell has only an s subshell, designated 1s; the second shell has both s and p subshells (2s and 2p); the third shell 3s, 3p, and 3d; and the fourth shell 4s, 4p, 4d, and 4f. (This pattern continues for higher-numbered shells, but this is enough for now.)

An s subshell is spherically symmetric and can hold a maximum of 2 electrons. A p subshell is dumbbell-shaped and holds 6 electrons, a d subshell 10 electrons, and an f subshell 14 electrons, with increasingly complicated shapes.

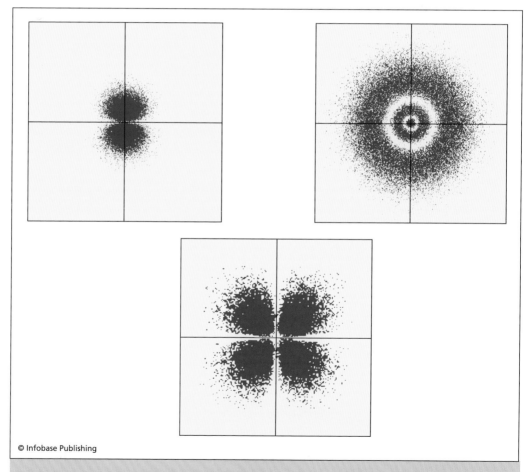

Some hydrogen wavefunction distributions for electrons in various excited states

As the number of electrons in an atom increases, so does the number of shells occupied by electrons. In addition, because electrons are all negatively charged and tend to repel each other *electrostatically,* as the number of the shell increases, the size of the shell increases, which means that electrons in higher-numbered shells are located, on the average, farther from the nucleus. Inner shells tend to be fully occupied with the maximum number of electrons they can hold. The electrons in the outermost shell, which is likely to be only partially occupied, will determine that atom's properties.

Physicists and chemists use *electronic configurations* to designate which subshells in an atom are occupied by electrons as well as how many electrons are in each subshell. For example, nitrogen is element number 7, so it has seven electrons. Nitrogen's electronic configuration is $1s^2 2s^2 2p^3$; a superscript designates the number of electrons that occupy a subshell. The first shell is fully occupied with its maximum of two electrons. The second shell can hold a maximum of eight electrons, but it is only partially occupied with just five electrons—two in the 2s subshell and three in the 2p. Those five outer electrons determine nitrogen's properties. For a heavy element like tin (Sn), electronic configurations can be quite complex. Tin's configuration is $1s^2 2s^2 2p^6 3s^2 3p^6 4s^2 3d^{10} 4p^6 5s^2 4d^{10} 5p^2$ but is more commonly written in the shorthand notation $[Kr]\, 5s^2 4d^{10} 5p^2$, where [Kr] represents the electron configuration pattern for the noble gas krypton. (The pattern continues in this way for shells with higher numbers.) The important thing to notice about tin's configuration is that all of the shells except the last one are fully occupied. The fifth shell can hold 32 electrons, but in tin there are only four electrons in the fifth shell. The outer electrons determine an element's properties. The table on page xxix illustrates the electronic configurations for nitrogen and tin.

ATOMS ARE HELD TOGETHER WITH CHEMICAL BONDS

Fundamentally, a chemical bond involves either the sharing of two electrons or the transfer of one or more electrons to form ions. Two atoms of nonmetals tend to share pairs of electrons in what is called a *covalent bond.* By sharing electrons, the atoms remain more or less electrically neutral. However, when an atom of a metal approaches an atom of a nonmetal, the more likely event is the transfer of one or more electrons from the metal atom to the nonmetal atom. The metal atom becomes

ELECTRONIC CONFIGURATIONS FOR NITROGEN AND TIN

ELECTRONIC CONFIGURATION OF NITROGEN (7 ELECTRONS)

Energy Level	Shell	Subshell	Number of Electrons
1	K	s	2
2	L	s	2
		p	3
			7

ELECTRONIC CONFIGURATION OF TIN (50 ELECTRONS)

Energy Level	Shell	Subshell	Number of Electrons
1	K	s	2
2	L	s	2
		p	6
3	M	s	2
		p	6
		d	10
4	N	s	2
		p	6
		d	10
5	O	s	2
		p	2
			50

a positively charged ion and the nonmetal atom becomes a negatively charged ion. The attraction between opposite charges provides the force that holds the atoms together in what is called an *ionic bond.* Many chemical bonds are also intermediate in nature between covalent and ionic bonds and have characteristics of both types of bonds.

IN CHEMICAL REACTIONS, ATOMS REARRANGE TO FORM NEW COMPOUNDS

When a substance undergoes a *physical change,* the substance's name does not change. What may change is its temperature, its length, its *physical state* (whether it is a solid, liquid, or gas), or some other characteristic, but it is still the same substance. On the other hand, when a substance undergoes a *chemical change,* its name changes; it is a different substance. For example, water can decompose into hydrogen gas and oxygen gas, each of which has substantially different properties from water, even though water is composed of hydrogen and oxygen atoms.

In chemical reactions, the atoms themselves are not changed. Elements (like hydrogen and oxygen) may combine to form compounds (like water), or compounds can be decomposed into their elements. The atoms in compounds can be rearranged to form new compounds whose names and properties are different from the original compounds. Chemical reactions are indicated by writing chemical equations such as the equation showing the decomposition of water into hydrogen and oxygen: $2 H_2O \ (l) \rightarrow 2 H_2 \ (g) + O_2 \ (g)$. The arrow indicates the direction in which the reaction proceeds. The reaction begins with the *reactants* on the left and ends with the *products* on the right. We sometimes designate the physical state of a reactant or product in parentheses—*s* for solid, *l* for liquid, *g* for gas, and *aq* for *aqueous* solution (in other words, a solution in which water is the solvent).

IN NUCLEAR REACTIONS THE NUCLEI OF ATOMS CHANGE

In ordinary chemical reactions, chemical bonds in the reactant species are broken, the atoms rearrange, and new chemical bonds are formed in the product species. These changes only affect an atom's electrons; there is no change to the nucleus. Hence there is no change in an element's identity. On the other hand, nuclear reactions refer to changes in an atom's nucleus (whether or not there are electrons attached). In most nuclear reactions, the number of protons in the nucleus changes, which means that elements are changed, or transmuted, into different elements. There are several ways in which *transmutation* can occur. Some transmutations occur naturally, while others only occur artificially in nuclear reactors or particle accelerators.

The most familiar form of transmutation is *radioactive decay,* a natural process in which a nucleus emits a small particle or *photon* of light. Three common modes of decay are labeled *alpha, beta,* and *gamma* (the first three letters of the Greek alphabet). Alpha decay occurs among elements at the heavy end of the periodic table, basically elements heavier than lead. An alpha particle is a nucleus of helium 4 and is symbolized as ^4_2He or α. An example of alpha decay occurs when uranium 238 emits an alpha particle and is changed into thorium 234 as in the following reaction: $^{238}_{92}U \rightarrow {}^4_2He + {}^{234}_{90}Th$. Notice that the parent isotope, U-238, has 92 protons, while the daughter isotope, Th-234, has only 90 protons. The decrease in the number of protons means a change in the identity of the element. The mass number also decreases.

Any element in the periodic table can undergo beta decay. A beta particle is an electron, commonly symbolized as β^- or e^-. An example of beta decay is the conversion of cobalt 60 into nickel 60 by the following reaction: $^{60}_{27}Co \rightarrow {}^{60}_{28}Ni + e^-$. The atomic number of the daughter isotope is one greater than that of the parent isotope, which maintains charge balance. The mass number, however, does not change.

In gamma decay, photons of light (symbolized by γ) are emitted. Gamma radiation is a high-energy form of light. Light carries neither mass nor charge, so the isotope undergoing decay does not change identity; it only changes its energy state.

Elements also are transmuted into other elements by nuclear *fission* and *fusion.* Fission is the breakup of very large nuclei (at least as heavy as uranium) into smaller nuclei, as in the fission of U-236 in the following reaction: $^{236}_{92}U \rightarrow {}^{94}_{36}Kr + {}^{139}_{56}Ba + 3n$, where n is the symbol for a neutron (charge = 0, mass number = +1). In fusion, nuclei combine to form larger nuclei, as in the fusion of hydrogen isotopes to make helium. Energy may also be released during both fission and fusion. These events may occur naturally—fusion is the process that powers the Sun and all other stars—or they may be made to occur artificially.

Elements can be transmuted artificially by bombarding heavy target nuclei with lighter projectile nuclei in reactors or accelerators. The transuranium elements have been produced that way. Curium, for example, can be made by bombarding plutonium with alpha particles. Because the projectile and target nuclei both carry positive charges, projectiles must be accelerated to velocities close to the speed of light to overcome the force of repulsion between them. The production of successively heavier nuclei requires more and more energy. Usually, only a few atoms at a time are produced.

ELEMENTS OCCUR WITH DIFFERENT RELATIVE ABUNDANCES

Hydrogen overwhelmingly is the most abundant element in the universe. Stars are composed mostly of hydrogen, followed by helium and only very small amounts of any other element. Relative abundances of elements can be expressed in parts per million, either by mass or by numbers of atoms.

On Earth, elements may be found in the lithosphere (the rocky, solid part of Earth), the hydrosphere (the aqueous, or watery, part of Earth), or the atmosphere. Elements such as the noble gases, the rare earths, and commercially valuable metals like silver and gold occur in only trace quantities. Others, like oxygen, silicon, aluminum, iron, calcium, sodium, hydrogen, sulfur, and carbon are abundant.

HOW NATURALLY OCCURRING ELEMENTS HAVE BEEN DISCOVERED

For the elements that occur on Earth, methods of discovery have been varied. Some elements—like copper, silver, gold, tin, and lead—have been known and used since ancient or even prehistoric times. The origins of their early metallurgy are unknown. Some elements, like phosphorus, were discovered during the Middle Ages by alchemists who recognized that some mineral had an unknown composition. Sometimes, as in the case of oxygen, the discovery was by accident. In other instances—as in the discoveries of the alkali metals, alkaline earths, and lanthanides—chemists had a fairly good idea of what they were looking for and were able to isolate and identify the elements quite deliberately.

To establish that a new element has been discovered, a sample of the element must be isolated in pure form and subjected to various chemical and physical tests. If the tests indicate properties unknown in any other element, it is a reasonable conclusion that a new element has been discovered. Sometimes there are hazards associated with isolating a substance whose properties are unknown. The new element could be toxic, or so reactive that it can explode, or extremely radioactive. During the course of history, attempts to isolate new elements or compounds have resulted in more than just a few deaths.

HOW NEW ELEMENTS ARE MADE

Some elements do not occur naturally, but can be synthesized. They can be produced in nuclear reactors, from collisions in particle accelerators, or can be part of the *fallout* from nuclear explosions. One of the elements most commonly made in nuclear reactors is technetium. Relatively large quantities are made every day for applications in nuclear medicine. Sometimes, the initial product made in an accelerator is a heavy element whose atoms have very short *half-lives* and undergo radioactive decay. When the atoms decay, atoms of elements lighter than the parent atoms are produced. By identifying the daughter atoms, scientists can work backward and correctly identify the parent atoms from which they came.

The major difficulty with synthesizing heavy elements is the number of protons in their nuclei (Z > 92). The large amount of positive charge makes the nuclei unstable so that they tend to disintegrate either by radioactive decay or *spontaneous fission*. Therefore, with the exception of a few transuranium elements like plutonium (Pu) and americium (Am), most artificial elements are made only a few atoms at a time and so far have no practical or commercial uses.

THE TRANSITION METALS SECTION OF THE PERIODIC TABLE

The metals in this book are called the transition metals and can be found in the central section of the periodic table in the appendix on page 137. They are introduced in the following groups: scandium and yttrium; titanium, zirconium, and hafnium; vanadium, niobium, and tantalum; chromium, molybdenum, and tungsten; manganese, technetium, and rhenium; iron, cobalt, and nickel; ruthenium, osmium, rhodium, iridium, palladium, and platinum; and copper, silver, and gold.

The book also includes zinc, cadmium, and mercury. Various authors classify these last three elements in different ways. Authors who adhere strictly to the definition of a transition metal as an element with an only partially occupied d-subshell will exclude zinc, cadmium, and mercury because their d-subshells are filled. However, in this book, as in books by some other authors, zinc, cadmium, and mercury are included with the transition metals because their properties are more similar to those of transition metals than to those of post-transition metals, whose properties are determined by partially filled p-subshells.

The following is the key to understanding each element's information box that appears at the beginning of each chapter.

Included in the transition metals are some of society's most industrially important substances. Iron, nickel, copper, gold, and silver are familiar metals that have played important roles in civilization for thousands of years. Titanium and hafnium are particularly important in today's economy—titanium as a structural material and hafnium as a new building block of modern electronics. In this book, readers will learn about the important properties of transition metals and how these elements are useful and possibly hazardous in everyday life.

Element		
K		M.P.°
L		B.P.°
M	E_Z	C.P.°
N		
O		
P	Oxidation states	
Q	Atomic weight	
	Abundance%	

Information box key. E represents the element's letter notation (for example, H = hydrogen), with the Z subscript indicating proton number. Orbital shell notations appear in the column on the left. For elements that are not naturally abundant, the mass number of the longest-lived isotope is given in brackets. The abundances (atomic %) are based on meteorite and solar wind data. The melting point (M.P.), boiling point (B.P.), and critical point (C.P.) temperatures are expressed in Celsius. Sublimation and critical temperatures are indicated by s and t.

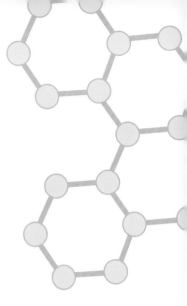

1

The Scandium Group

The scandium group consists of scandium (Sc, element 21) and yttrium (Y, element 39). Scandium is the first metal besides aluminum to exist in compounds almost exclusively in the "+3" oxidation state. The stories of the discoveries of scandium and the elements in its group, their properties, and their chemistry are described in this chapter.

From hydrogen (element 1) through calcium (element 20), all of the electrons in ground-state atoms are located in either *s* or *p* subshells. An important change occurs with scandium: For the first time, a *d* subshell is occupied by an electron. Scandium, yttrium, and lanthanum each mark the beginning of a series of transition elements. Scandium's outermost electron is the first electron in the 3d subshell, yttrium's outermost electron is the first electron in the 4d subshell, and lanthanum's

THE BASICS OF SCANDIUM

Symbol: Sc
Atomic number: 21
Atomic mass: 44.9559
Electronic configuration: $[Ar]4s^2 3d^1$

T_{melt} = 2,806°F (1,541°C)
T_{boil} = 5,137°F (2,836°C)

Abundance in Earth's crust = 26 ppm

		Scandium	
2			1541°
8		**Sc**$_{21}$	2836°
9			
2			
		+3	
		44.955910	
		1.12×10^{-7}%	

Isotope	Z	N	Relative Abundance
$^{45}_{21}$Sc	21	24	100%

THE BASICS OF YTTRIUM

Symbol: Y
Atomic number: 39
Atomic mass: 88.9058
Electronic configuration: $[Kr]5s^1 4d^1$

T_{melt} = 2,772°F (1,522°C)
T_{boil} = 6,053°F (3,345°C)

Abundance in Earth's crust = 29 ppm

		Yttrium	
2			1522°
8		**Y** 39	3345°
18			
9			
2			
		+3	
		88.90585	
		1.51×10^{-8}%	

Isotope	Z	N	Relative Abundance
$^{89}_{39}$Y	39	50	100%

is the first electron in the 5d subshell. The combination of two valence electrons in an *s* subshell plus the one electron in a *d* subshell gives each of these elements a total of three valence electrons. Therefore, it is not surprising that the valence state in compounds of these elements is exclusively "+3".

Because of the change in type of valency in going from the group IIA (alkaline earth) elements to the group IIIA (scandium family) elements, there is a marked change in properties. The scandium family of elements is characterized by significantly higher melting points and boiling points than the alkaline earths, as well as higher densities, greater hardness, decreasing strength of the oxides as bases, and higher *heats of fusion* and *vaporization*.

THE ASTROPHYSICS OF SCANDIUM AND YTTRIUM

Scandium and yttrium nuclei are not built in the cores of average stars like the Sun. Instead, these elements build up over thousands of years in the atmospheres of more massive stars via *neutron capture*. Neutrons become available from the capture of alpha particles by ^{13}C and ^{22}Ne as follows:

$$^{13}_{6}C + ^{4}_{2}\alpha \rightarrow ^{16}_{8}O + ^{1}_{0}n$$

$$^{22}_{10}Ne + ^{4}_{2}\alpha \rightarrow ^{25}_{12}Mg + ^{1}_{0}n$$

Because the synthesis proceeds slowly due to a low density of neutrons, it is called the slow process or *s-process*.

Both of these elements can also be synthesized in *supernovae* via the rapid capture by iron nuclei of a succession of neutrons, which is called the *r-process*. The scandium and yttrium produced by either process then flows into the surrounding atmosphere on the stellar wind or is blasted into space by the supernova explosion and mixes with the *interstellar medium* (ISM). As stars like the Sun form, they collect the surrounding gases and incorporate a small portion of these elements into their makeup.

A puzzling phenomenon is the variability of scandium abundance in various types of stars. Some *chemically peculiar stars* have been particularly of interest in this regard. Stars of *spectral type* A are about one and a half to three times as massive as the Sun and have much hotter surface temperatures. Their spectrum is typically dominated by hydrogen lines, with somewhat dimmer iron and calcium lines. They also have a rather rapid rotation rate. Some type A stars—especially those in *binary systems*—rotate much more slowly, however, probably because of the gravitational influence of the companion star. In these stars, more metallic elements are observed, but the scandium content is much less than expected. One possibility for the discrepancy may be that the slow rotation limits large-scale mixing of elements; scandium may exist in higher abundance deep within the star.

Another peculiar A star, the so-called manganese-scandium star, "φ Herculis," displays normal abundances of magnesium and silicon, but extreme overabundances of scandium (25 times the norm) and yttrium (thousands of times the norm). Yttrium is, however, overly deficient in metal-poor stars.

These dilemmas will not be resolved without further investigation by astrophysicists.

DISCOVERY AND NAMING OF SCANDIUM AND YTTRIUM

In the 1830s, a new mineral was discovered in Scandinavia. The German metallurgist C. J. A. Theodor Scheerer (1813–75) named the mineral *euxenite* from the Greek word *euxenos,* which means "kind to strangers." Scheerer chose that name because euxenite contained a number of rare, or "strange," elements, among them tantalum, titanium, yttrium, uranium, cerium, and lanthanum.

When Dmitri Mendeleev published his periodic table of the elements in 1869, he predicted the existence of at least three as-yet-undiscovered elements that would fill empty spaces in the table. Two of those elements were gallium and germanium. The third element was scandium. Mendeleev called this undiscovered element *eka-boron* because he felt it would resemble boron.

In 1879, the Swedish analytical chemist Lars Fredrik Nilson (1840–99), working in the laboratory of the Swedish chemist Jöns Jacob Berzelius (1779–1848), also analyzed euxenite and discovered that it contained an unknown substance. Upon careful investigation of the unknown substance, Nilson concluded that its properties matched the properties of eka-boron that had been predicted by Mendeleev (see the table on page 5).

Euxenite had not been found anywhere other than in Scandinavia. In addition, Nilson's new element had been found in the mineral gadolinite, which at the time was only known to occur in Scandinavia. Con-

COMPARISON OF EKA-BORON AND SCANDIUM

	PROPERTIES MENDELEEV PREDICTED FOR EKA-BORON (EB)	PROPERTIES NILSON FOUND FOR SCANDIUM (SC)
Atomic mass	44	44
Formula of oxide	Eb_2O_3	Sc_2O_3
Density of oxide	3.5	3.86
Color of salts	Colorless	Colorless
Solubility of carbonate	Insoluble	Insoluble
Formula of chloride	$EbCl_3$	$ScCl_3$

sequently, Nilson decided to name the new element *scandium* in honor of Scandinavia.

Scandium emits no lines in the visible region of the electromagnetic spectrum. Nevertheless, scandium can be detected by its lines in other regions of the spectrum. Until the beginning of the 20th century, scandium was thought to be one of the rarest elements on Earth. In 1908, however, the British chemist Sir William Crookes (1832–1919) discovered that scandium actually is widely distributed on Earth. Scandium is often recovered from uranium *tailings*.

Until the beginning of the 20th century, scandium was thought to be one of the rarest elements on Earth. *(American Elements)*

Yttrium was first found in minerals near the Swedish town of Ytterby. The minerals also contained the rare earth elements ytterbium, erbium, and terbium. All four names were derived from the name of the town. Credit for yttrium's discovery in 1794 is given to the Finnish chemist Johan Gadolin (1760–1852). Gadolin served as a professor of chemistry at the University of Åbo, where he isolated a number of elements. In recognition of Gadolin's contributions, chemists named element 64—which he had discovered—*gadolinium* in his honor.

THE CHEMISTRY OF SCANDIUM AND YTTRIUM

Scandium is element number 21 and ranks 35th in abundance. It is almost always found in association with yttrium and with the lanthanides. Scandium is a soft, silvery metal with a density of 3.0 g/cm^3. Although trace quantities (measured in parts per billion) can be measured in human blood and tissues, scandium has no known biological function.

Scandium is a fairly reactive element. It dissolves in water and most acids to form the Sc^{3+} ion. It reacts with atmospheric oxygen at normal temperatures. In nature, scandium easily forms an insoluble hydroxide—$Sc(OH)_3$. In the laboratory, a *hydride* can be made. Surprisingly, it has a formula of ScH_2, suggesting that the scandium ion is only Sc^{2+}. It is not understood why scandium is not "+3" in the hydride.

The scandium ion tends to form compounds with most common *univalent* anions (ions with charges of "–1"). Thus, $Sc(NO_3)_3$, ScF_3, $ScCl_3$, $ScBr_3$, and ScI_3 are easily formed. These compounds are all white in color and are all soluble in dilute aqueous solution.

Yttrium is element number 39 and the 43rd most abundant element on Earth. Although most people have probably never even heard of yttrium, it is actually twice as common as lead. Most yttrium today comes from Australia. The primary ytrrium-containing mineral is called xenotime (yttrium phosphate, YPO_4). Another source of yttrium is uranium mines in Canada.

Yttrium has a density of 4.5 g/cm^3. Metallic yttrium—along with the rare earth elements with which it is found—usually is produced by *electrolysis* of the chloride salt, as shown in the following chemical reaction:

$$2 \; YCl_3 \; (l) \rightarrow 2 \; Y \; (s) + 3 \; Cl_2 \; (g).$$

Yttrium has a silvery-metallic luster and is found in most rare-earth minerals. *(Charles D. Winters/Photo Researchers, Inc.)*

All of the elements found with yttrium are so similar, however, that the yttrium metal formed by electrolysis tends not to be completely pure. The best method for separating yttrium ions from other closely related ions is by *ion exchange,* in which a solution containing a mixture of the various ions is poured through a resin. The different ions will travel through the resin at different speeds that depend on the relative charges and sizes of the ions. As the ions pass through the resin, the individual ions can be collected. This process is referred to as *elution.*

Yttrium's chemistry is dominated by the "+3" oxidation state. At high temperatures, metallic yttrium *oxidizes* to Y_2O_3. Only the nitrate—$Y(NO_3)_3$—is soluble in water. Other common yttrium compounds are $Y(OH)_3$, $Y_2(CO_3)_3$, $Y_2(C_2O_4)_3$, YF_3, and YPO_4. Yttrium hydroxide is yellow in color and is slightly soluble in water. The other compounds are white in color and insoluble.

THE RARE EARTH CONNECTION

In the first two horizontal rows of the *d* block, the progressive filling of 3d and 4d subshells, respectively, continues across the periods—although not completely regularly, as will be shown later. At the beginning of the

SCANDIUM AND ESPIONAGE

While radioactive elements can be beneficial as diagnostics in the human body and are particularly useful for the *imaging* and treatment of tumors, the use of radioactive sources as *tracers* also has its nefarious side. Since radioactivity is invisible to the eye, but detectable by other means (such as *Geiger counters*), it is perhaps not surprising that elements like scandium 46 have served a purpose in espionage.

It is well documented that the East German secret police—the Ministry for State Security, or "Stasi"—employed scandium 46 for surveillance of human subjects during the 1970s and 1980s. Scandium 46 decays by emission of a high-energy electron that can be easily detected several yards (several meters) from the source, even if there is an intervening barrier such as a wall. Anyone having on his or her person a scandium 46 source could therefore be "seen" without seeing the observer. It is not a naturally occurring isotope, however, so people would not normally carry around ^{46}Sc, and the Stasi went to great efforts to plant sources on their surveillance targets. The radioactive material was acquired from a German nuclear research center, and special liquid sprays were developed that could be applied to cash or to a person's clothing or injected into a ballpoint pen.

Hundreds of subjects were trailed in this manner with little thought given to possibly deleterious health effects. Since the half-life of ^{46}Sc is 83 days, a person carrying an irradiated object for a couple of months would be at risk for cellular damage and *mutation*. High-energy electrons are *ionizing particles*. That is, upon contact they can remove or add electrons to a cell, which can trigger *carcinogenic* effects. No known data have been collected regarding the effects on these subjects. While the practice ended in East Germany upon the dismantling of the Berlin Wall, it is conceivable that this insidious method of surveillance continues today in other countries.

third row, however, an abrupt change occurs. After lanthanum, instead of continuing to fill the 5d subshell, the additional valence electrons of the next 14 elements fill the 4f subshell—marking the first time in the periodic table that an *f* subshell is populated with electrons in the ground state of an atom. Because the 4f subshell is in an energy level that is "inside" the energy level of the 5d subshell, these 14 elements are considered to constitute a row of so-called inner transition elements. The sizes of their atoms and ions are all so similar that these elements exhibit virtually identical chemical and physical properties. Thus, chemists think of them as effectively occupying the same place as lanthanum in the periodic table, and they are referred to as the *lanthanide series*.

It should be noted that there is a considerable jump in atomic number in going from lanthanum (element number 57) to hafnium (element 72), the element immediately following lanthanum in the periodic table. The jump from element 57 to element 72 results from the 14 lanthanides that have been separated out of the main part of the table. Beginning with hafnium, the trend is once again to progressively fill the 5d subshell with valence electrons as one moves horizontally across the period.

A similar situation occurs following actinium (element number 89) in the table. In actinium, one valence electron populates the 6d subshell. With the next 14 elements, however, the 5f subshell fills instead of the 6d. Therefore, those 14 elements—the so-called actinides—are also separated from the main part of the periodic table and constitute a horizontal row of elements that is located directly beneath the lanthanides. Together, the lanthanides and actinides are referred to as inner transition elements, or rare earths. After the last actinide—lawrencium, element 103—the filling of the 6d subshell resumes with rutherfordium (element 104) and continues through the last *period* of transition elements that have been discovered to date. (All of these elements—beginning with rutherfordium—do not exist naturally on Earth, but are produced in particle accelerators.)

SCANDIUM STRENGTHENS ALUMINUM ALLOYS

For improving the strength of aluminum materials, there is no better single element to add than scandium. Scandium-aluminum alloys are strong mainly because grain size is reduced. Small grains embedded in a metal *lattice* inhibit the transfer of a disturbance (such as caused by a forceful blow) through the material. Dislocation of the lattice is impeded at grain boundaries, and vibration is dispersed better when very small grains are distributed throughout the material.

Sc-Al alloys also exhibit excellent *thermal stability*. This feature makes a high-strength aluminum alloy easier to weld. It also reduces *recrystallization*. Crystallization weakens any object by restoring a rigid lattice that can have a higher *shear* tendency, leading to loss of strength. This tends to occur in items like bicycle frames and baseball bats that are *cold-worked* in their manufacture. Using Sc-Al alloys can alleviate this problem.

Recent studies have shown that the addition of small amounts of zirconium to scandium-aluminum alloys can have a dramatic effect on strength and thermal stability. This effect is discussed in chapter 2.

TECHNOLOGY AND CURRENT USES OF THE SCANDIUM GROUP

Scandium has relatively few uses. Some scandium, however, is added to aluminum and magnesium alloys to strengthen them. Theoretically, scandium could replace aluminum as a structural material. Scandium weighs the same as aluminum but has a much higher melting point. However, the much higher cost of scandium compared to aluminum makes large-scale use of scandium economically prohibitive.

Mixed with molybdenum, scandium helps to inhibit the corrosion of zirconium. Scandium metal has been used as a filter of high-speed neutrons in nuclear reactors. It has also historically been used as a tracer in East German espionage.

The largest commercial use of scandium—in the form of scandium iodide—is in high-intensity lighting. A mixture of scandium carbide and titanium carbide produces a *ceramic* material second in hardness only to diamond.

Yttrium has a number of uses. Yttrium is used in the red *phosphor* in television picture tubes and flat plasma screens and in cobalt-based alloys. Yttrium oxide is used in neodymium-*doped* laser crystals and is a component of certain *superconducting* alloys. Yttrium iron *garnet* has applications in optics, acoustics, and radar.

Yttrium-aluminum-garnet crystals are used in some lasers. *(Peter Ginter/Getty Images)*

Yttrium's radioactive isotope, ^{90}Y, is produced by beta decay of ^{90}Sr and is used in *nuclear medicine,* where it has proven particularly useful in treating tumors of the liver. Given that yttrium is a relatively abundant and inexpensive element, it may be expected that additional applications will be found.

2

The Titanium and Vanadium Groups

The titanium group of elements consists of titanium (Ti, element 22), zirconium (Zr, element 40), and hafnium (Hf, element 72). The vanadium group consists of vanadium (V, element 23), niobium (Nb, element 41), and tantalum (Ta, element 73).

The titanium metals are found in group IVA. Each of these elements is *tetravalent,* meaning that its chemistry is dominated by the "+4" oxidation state. Titanium is the best-known element in the group. Titanium oxide (TiO_2) is a white paint pigment. Titanium alloys—such as those used in golf clubs, surgical instruments, and *prosthetics*—are known for their exceptional strength. *Zircons* are familiar diamond-like gems composed of zirconium silicate, $ZrSiO_4$. Hafnium is probably less familiar to most people, but it is an important component of control rods in the fuel assemblies of nuclear reactors and is important in the semiconductor industry.

THE BASICS OF TITANIUM

Symbol: Ti
Atomic number: 22
Atomic mass: 47.867
Electronic configuration: $[Ar]4s^23d^2$

T_{melt} = 3,034°F (1,668°C)
T_{boil} = 5,949°F (3,287°C)

Abundance in Earth's crust = 6600 ppm

Isotope	Z	N	Relative Abundance
$^{46}_{22}$Ti	22	24	8.25%
$^{47}_{22}$Ti	22	25	7.44%
$^{48}_{22}$Ti	22	26	73.72%
$^{49}_{22}$Ti	22	27	5.41%
$^{50}_{22}$Ti	22	28	5.18%

Titanium		
2	**Ti**$_{22}$	1668°
8		3287°
10		
2		
	+2 +3 +4	
	47.867	
	7.8×10^{-6}%	

THE BASICS OF ZIRCONIUM

Symbol: Zr
Atomic number: 40
Atomic mass: 91.224
Electronic configuration: $[Kr]5s^24d^2$

T_{melt} = 3,371°F (1,855°C)
T_{boil} = 7,968°F (4,409°C)

Abundance in Earth's crust = 130 ppm

Isotope	Z	N	Relative Abundance
$^{90}_{40}$Zr	40	50	51.45%
$^{91}_{40}$Zr	40	51	11.22%
$^{92}_{40}$Zr	40	52	17.15%
$^{94}_{40}$Zr	40	54	17.38%
$^{96}_{40}$Zr	40	56	2.80%

Zirconium		
2	**Zr**$_{40}$	1855°
8		4409°
18		
10		
2		
	+4	
	91.224	
	3.72×10^{-8}%	

THE BASICS OF HAFNIUM

Symbol: Hf
Atomic number: 72
Atomic mass: 178.49
Electronic configuration:
[Xe]$6s^2 4f^{14} 5d^2$

T_{melt} = 4,051°F (2,233°C)
T_{boil} = 8,317°F (4,603°C)

Abundance in Earth's crust = 3.3 ppm

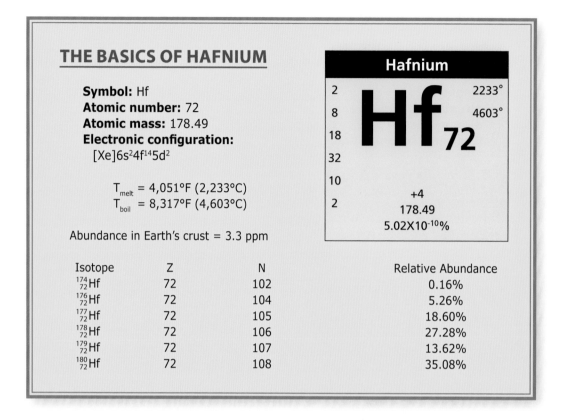

			Hafnium		
2					2233°
8		**Hf**			4603°
18			72		
32					
10					
2			+4		
			178.49		
			5.02X10^{-10}%		

Isotope	Z	N	Relative Abundance
$^{174}_{72}$Hf	72	102	0.16%
$^{176}_{72}$Hf	72	104	5.26%
$^{177}_{72}$Hf	72	105	18.60%
$^{178}_{72}$Hf	72	106	27.28%
$^{179}_{72}$Hf	72	107	13.62%
$^{180}_{72}$Hf	72	108	35.08%

The vanadium metals are found in group VA. Each of these elements is *pentavalent,* meaning that the "+5" oxidation state is important to each. Vanadium, however, forms compounds that exhibit a relatively large number of oxidation states. In addition to the "+5" state, there are the "+2," "+3," and "+4" states. Similarly, niobium exhibits a "+3" state in addition to the "+5".

THE ASTROPHYSICS OF THE TITANIUM GROUP:
Ti, Zr, Hf

Titanium 44, an α-process element, is synthesized in very large supernovae. *Alpha-process* elements are synthesized by sequential absorption of alpha particles, which consist of four particles, so these elements all have mass numbers that are multiples of four. The following reaction synthesizes titanium 44 in the atmospheres of expanding supernovae, which distribute this radioactive isotope into the surrounding space.

$$^{40}_{20}\text{Ca} + ^{4}_{2}\alpha \rightarrow ^{44}_{22}\text{Ti}$$

Titanium 44 decays with a half-life of 59 days to ^{44}Sc. Scandium 44 is itself unstable and decays with a half-life of 1.8 days via emission of a photon with the energy of 1.156 million *electron volts* (MeV), producing the stable isotope ^{44}Ca. This is the signature emission line in a stellar spectrum that tells astronomers ^{44}Ti was once present.

A curious stellar object is the so-called titanium star, *Cas A,* which is a supernova remnant about 9,000 light years away with a *neutron star* at its center. The blast should have been observable on Earth in the mid-17th century, although there is no record of any such viewing. This supernova is unique in that it produced an unprecedented excess of ^{44}Ti relative to ^{56}Ni (which is produced in all supernovae). Since the synthesis of titanium requires temperatures on the order of 5 billion Kelvin, and gravitational contraction temperatures rise directly with the mass of a star, the original star must have been an exceptionally massive object before its demise.

The most abundant isotope of titanium on Earth—Ti-48, which is stable—is not a product of stellar nucleosynthesis, but of the beta decay of terrestrial scandium 48 or vanadium 48.

Like yttrium and most other elements heavier than iron, zirconium and hafnium syntheses occur in stars via neutron capture onto lighter elements. Zirconium abundance in stellar atmospheres presents some surprises, however. Like yttrium, it appears to be deficient in stars with low metal content, but the ratios of these two elements relative to iron fluctuate from star to star, even when the stars are of similar types. However, there is less variability in abundances of these elements relative to titanium, which is puzzling.

Another area of interest is the so-called zirconium conflict, which refers to a confusing situation observed in some *HgMn stars* where the spectral line corresponding to doubly ionized zirconium (Zr^{2+}) is much stronger than that belonging to the singly ionized atoms (Zr^+). This is counterintuitive, because more energy is required to remove two electrons than just one. A broadening of excitation energy levels within the atom might, however, cause an atomic emission line to appear dimmer. One way this could happen is through bombardment of zirconium atoms by free-flying electrons, of which huge numbers exist in the hot gas of a star. This electron-impact broadening would have a greater influence on Zr^+ than on Zr^{2+}, but the calculated effect is not enough to explain the observed differences.

The abovementioned anomalies indicate an incomplete scientific understanding of the dynamic processes within stellar atmospheres and interiors and the need for more observational data.

DISCOVERY AND NAMING OF TITANIUM, ZIRCONIUM, AND HAFNIUM

Titanium was discovered by the English clergyman, the Reverend William Gregor (1761–1817). A graduate of Cambridge University, Gregor served several churches but spent most of his career as the rector at a church in Creed, England, from 1793 until 1817. Gregor's friend, chemist John Warltire (1739–1810), introduced him to chemistry and mineralogy. Gregor was particularly attracted to mineralogy and was well known for his analyses of England's minerals. Intrigued by a black magnetic sand from his parish, he analyzed it and found it to be 46.56 percent *magnetite* (an iron oxide), 3.5 percent *silica*, 45 percent an unknown reddish-brown substance, and 4.94 percent other material.

Gregor showed his results to another friend, John Hawkins. The two men agreed that the reddish-brown substance was a mineral that most likely contained a new element. Hawkins suggested naming the new element *menachanite* after the Menachan Valley in which the sand had been found. Gregor's parish responsibilities prevented him from pursuing the matter further. Unfortunately, he died of tuberculosis in 1817 without ever returning to his research.

Because of Gregor's declining health, menachanite was all but forgotten. In 1795, however, the German chemist Martin Heinrich Klaproth (1743–1817) began investigating a specimen of the mineral rutile (titanium oxide) from Hungary. From the rutile, he separated a metallic oxide whose properties remarkably resembled the properties of menachanite. Klaproth began studying both minerals, carefully comparing their properties, and concluded that they were the same metallic oxide. Although he gave William Gregor full credit for priority of discovery, Klaproth chose not to adopt Gregor's name of menachanite for the new element. Instead, deciding there were no special properties of the element or peculiarities regarding its origin, he chose a name that had nothing to do with the element's properties. Klaproth chose the name *titanium* after the Greek gods called the *Titans,* the children of Uranus and Gaia.

A number of minerals contain zirconium. Zircon, a trace mineral common to most *granites,* has been used as a gemstone since ancient times. Until nearly 1800, however, all analyses of zirconium minerals were erroneous. They were reported to contain silica, iron oxide, alumina, lime (calcium oxide), and other minerals, but nothing that would have been a new element.

In 1789, Klaproth analyzed zircon and discovered that it contained the mineral zirconia (later shown to be zirconium oxide, ZrO_2). In 1824, Berzelius heated a sample of zirconia with potassium metal. Potassium reduced the zirconium to an impure powdered form of the metal. During the next 90 years, chemists improved the process for isolating zirconium and gradually succeeded in obtaining samples of zirconium of successively higher purity. In 1914, a completely pure sample of zirconium was finally obtained by *reducing* zirconium tetrachloride ($ZrCl_4$) with sodium. In the end, Klaproth was credited with zirconium's discovery. The name *zirconium* itself was derived from the mineral zirconia.

In 1911, an element was discovered that was believed to be a lanthanide, which would have placed its atomic number between 57 and 71. After World War I, this element was found to occur mostly in titanium ores and to be more similar to zirconium than to the lanthanides. The Danish physicist Niels Bohr (1885–1962) suggested that this unknown element was more likely a transition metal in the titanium family. In 1923, acting upon Bohr's suggestion, the Hungarian chemist George Charles de Hevesy (1889–1966) and his coworker, the German physicist Dirk Coster (1889–1950), used *X-ray analysis* to prove that the atomic number of the unknown element had to be 72, which placed the element after the lanthanide series and below zirconium. Although neither de Hevesy nor Coster was Danish, the two men decided to name element 72 *hafnium,* after Bohr's home of Copenhagen, Denmark.

THE CHEMISTRY OF THE TITANIUM GROUP

Titanium is element 22, with a density of $4.5 g/cm^3$. Titanium is a silvery-white metal that is lighter and stronger than steel and very corrosion resistant. The ninth most abundant element in Earth's crust— and the fourth most abundant metal—titanium is widely distributed. The most important titanium ores are the *igneous* minerals ilmenite ($FeTiO_3$), which has black or brownish-black opaque crystals, and rutile (TiO_2), which can be reddish-brown, red, yellow, or black in color. Brookite and anatase are similar in composition to rutile but occur mostly in *metamorphic* rocks and have different crystalline structures than rutile does. Most ilmenite in the United States is mined in the state of New York, but the quantity of imported titanium exceeds domestic production.

Zirconium is element 40, with a density of 6.5 g/cm^3. Zirconium exhibits the same strength and corrosion resistance as titanium, and, in addition, is resistant to attack by concentrated hydrochloric, nitric, and sulfuric acids. Because of its high resistance to corrosion, zirconium is used in nuclear reactors. Pure zirconium is soft, *ductile,* and *malleable,* but just a 1 percent impurity renders it hard and brittle. Zirconium is the 18th most abundant element in Earth's crust, but is not widely distributed. It is obtained principally from two minerals: baddeleyite (ZrO_2) and zircon ($ZrSiO_4$ or $ZrO \cdot SiO_2$), a gem that can range from transparent to opaque, and which can be colorless, red, brown, yellow, green, or gray in color. Zircon's crystals can be an inch or more across, with especially large ones found in deposits in Canada and Australia. About 10 percent of the zirconium used in the United States comes from zircon mines in North Carolina, New Jersey, New York, and Pennsylvania. The rest is imported largely from Brazil and Australia, where zircon is recovered from beach sand.

Hafnium is element 72, with a relatively high density of 13.1 g/cm^3. It is the 45th most abundant element in Earth's crust. Hafnium has few minerals of its own, but tends to constitute about 1 to 2 percent of zirconium minerals. In addition, there are rarer minerals in which the hafnium content exceeds the zirconium content. Examples include alvite (composed of beryllium, hafnium, thorium, yttrium, zirconium, and silicon oxides) and cyrtolite (composed of hafnium, zirconium, and silicon oxides). The annual production of hafnium is relatively low for a metal.

All three metals are difficult to obtain from their ores. Titanium, for example, is obtained by mixing TiO_2 with carbon in a furnace. This process yields both titanium metal and titanium carbide (TiC), from which the pure metal has to be separated. Both titanium and zirconium can also be obtained by reducing their chlorides ($TiCl_4$ and $ZrCl_4$) with magnesium metal at high temperatures, as shown for titanium in the following chemical equation:

$$TiCl_4 \ (l) + 2 \ Mg \ (l) \rightarrow Ti \ (l) + 2 \ MgCl_2 \ (l).$$

When the mixture is cooled sufficiently for solid titanium to form, magnesium chloride's solubility in water allows the magnesium chloride to be washed away, leaving solid titanium metal behind.

These elements exist mostly in the "+4" state in chemical compounds. Of the three elements, titanium alone displays a limited chemistry in the "+2" and "+3" oxidation states. In the "+2" state, titanium manifests ionic bonding. In the "+3" and "+4" states, the chemical bonding exhibited by these elements is largely covalent. The table below lists some of the common compounds in which these elements are in the "+4" oxidation state and shows some of their similarities and differences.

Most of these compounds are white in color. Where they exist, the chlorides, nitrates, and sulfates are soluble in dilute aqueous solution. The oxides and sulfides are insoluble.

In the "+2" oxidation state, titanium forms titanium oxide (TiO) and titanium dichloride ($TiCl_2$). TiO is an ionic compound similar in properties to the oxides of the alkaline earths (CaO or MgO, for example). Both TiO and $TiCl_2$ are powerful reducing agents and react readily

COMMON COMPOUNDS OF THE TITANIUM GROUP

	Cl^-	NO_3^-	O^{2-}	CO_3^{2-}	SO_3^{2-}	S^{2-}
Ti	$TiCl_4$	None	TiO_2	None	None	TiS_2
Zr	$ZrCl_4$	$Zr(NO_3)_4$	ZrO_2	None	$Zr(SO_4)_2$	None
Hf	$HfCl_4$	None	HfO_2	None	None	None

with water, reducing the hydrogen in water to hydrogen gas, as shown in the following chemical equations:

$$TiO\ (s) + H_2O\ (l) \rightarrow TiO_2\ (s) + H_2\ (g);$$

$$TiCl_2\ (aq) + 2\ H_2O\ (l) \rightarrow TiO_2\ (s) + 2\ HCl\ (aq) + H_2\ (g).$$

Because Ti^{2+} reacts so readily with water, it is unstable in water and exhibits no other chemistry in aqueous solution.

In the "+3" oxidation state, titanium exists as the titanous ion, Ti^{3+}, which is violet in color. Ti^{3+} forms titanium *sesquioxide* (Ti_2O_3), titanous hydroxide ($Ti[OH]_3$), and titanous trichloride ($TiCl_3$). Ti_2O_3 is completely insoluble in water and can be produced by reducing TiO_2 with hydrogen gas, as shown by the following chemical equation:

$$2\ TiO_2\ (s) + H_2\ (g) \rightarrow Ti_2O_3\ (s) + H_2O\ (l).$$

Titanous hydroxide is a powerful reducing agent because of its tendency to be oxidized to the "+4" state. It will also decompose into TiO_2 and hydrogen gas, as shown by the following reaction:

$$2\ Ti(OH)_3\ (aq) \rightarrow 2\ TiO_2\ (s) + 2\ H_2O\ (l) + H_2\ (g).$$

Pure titanic oxide (TiO_2, also called "titanium dioxide") is white and has been used to replace white lead pigments in paints because of lead's toxicity. Titanic chloride ($TiCl_4$, also called titanium tetrachloride) is used to produce smoke screens, the smoke consisting mostly of $TiCl_4 \cdot 5H_2O$. Titanium also forms negatively charged *oxyanions* called *titanates*. Titanates can contain anions in the form of TiO_3^{2-} or TiO_4^{2-}. In both cases, titanium is in the "+4" oxidation state. Examples of titanate compounds include $CaTiO_3$, $Ba_2Ti_3O_8$, $ZnTiO_3$, $PbTiO_3$, and K_2TiO_3.

The titanium chlorides—$TiCl_2$, $TiCl_3$, and $TiCl_4$—demonstrate a general trend in bonding and properties that tends to be true in general of transition metal halides that exhibit increasing oxidation state of the metal. $TiCl_2$ and $TiCl_3$ are ionic crystals at room temperature with relatively high melting points. On the other hand, $TiCl_4$ is a covalently bonded molecular species that is a liquid at room temperature and that boils at 137°C.

The chemistry of zirconium and hafnium is so similar that, at one time, mixtures of the two were thought to be a single element. In analogy to titanium, zirconium forms *zirconates,* mostly of the form ZrO_3^{2-}. Examples include Na_2ZrO_3, $CaZrO_3$, and $PbZrO_3$. Also in analogy to titanium, zirconium forms a tetrachloride, $ZrCl_4$, and zirconyl chloride, $ZrOCl_2$. Several *fluorozirconates* are known. Examples include K_2ZrF_6 and $BaZrF_6$. The silicate, $ZrSiO_4$, which occurs in the mineral zircon, is a gem and can substitute for diamonds. In all of these cases, zirconium is in the "+4" oxidation state.

Hafnium's chemistry is virtually identical to that of zirconium, which explains why hafnium is found as a constituent of all zirconium minerals. Hafnium and zirconium are extremely difficult to separate from each other. Separation is best achieved using *ion exchange chromatography.*

TITANIUM IN THE AEROSPACE INDUSTRY

Titanium's mechanical properties make it an excellent structural material, particularly for the aerospace industry. Its high strength-to-weight ratio means that lighter aircraft can be built with better durability than aluminum or even steel. Its low *thermal conductivity* and structural stability at high temperatures—up to 1,500°F (815°C)—means that titanium and its alloys are ideal for spacecraft and jet engines. The element has been used in space capsules since the Mercury *Freedom 7* in 1961. Perhaps the most famous aircraft made with titanium was the SR-71 *Blackbird,* the fastest jet airplane in existence (though now retired). With a maximum design airspeed of 2,500 mph (4,023 km/h), the frictional heat required the airframe be able to withstand temperatures up to 500°F (250°C) without deformation or crystallization, making titanium the material of choice.

Titanium's crystal structure depends on the *ambient* temperature. From room temperature up to 1,621°F (883°C), pure titanium exists in a hexagonal, close-packed (HCP) crystal structure (α-phase). At higher temperatures, it transforms to a body-centered cubic (BCC) crystal structure (β-phase). Titanium alloys are produced with internal structure considerations. The structure can be modified to a distinct mix of α- and β-phases using stabilizers. Aluminum, oxygen, gallium, germanium, calcium, and nitrogen are α-phase stabilizers when alloyed

Titanium's mechanical properties make it an excellent structural material for building jet engines. *(Steve Mann/ Shutterstock)*

with titanium, and can help maintain the HCP structure. Vanadium and molybdenum are β-phase stabilizers, allowing for a mixture of crystal structures below the transition temperature. Zirconium and tin are known to help strengthen a titanium alloy.

Not only aircraft manufacturers, but makers of sports equipment, automobiles, pressure vessels, and many others would like to incorpo-

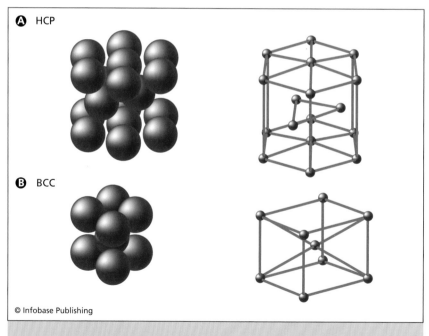

Ⓐ HCP

Ⓑ BCC

© Infobase Publishing

Depending on the temperature, titanium's crystal structure may take on hexagonal close-packed (HCP) or body-centered cubic (BCC) geometry.

rate more titanium into their designs. Unfortunately, the element in its pure form is difficult and expensive to extract, process, and fabricate. For an identical end product, an order-of-magnitude increase in cost is typical when compared with stainless steel. Researchers at the U.S. Department of Defense, among others, are investigating ways to reduce the production expense.

ZIRCONIUM IN ALUMINUM ALLOYS

As explained in chapter 1, adding scandium to aluminum strengthens the material because the size of grains embedded in the aluminum lattice can be reduced. The process by which the small grains form, however, has been something of a mystery until only very recently, when

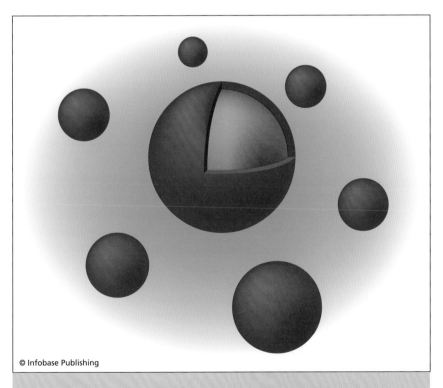

© Infobase Publishing

Simplified schematic representation of core-shell precipitates in Al-Sc-Zr alloys: Blue represents the Zr-rich shell and red represents the Sc-rich core. The precipitates are suspended in a sea of Al-rich alloy (green). *(Source: Nature Materials 5 (June 2006): 435)*

researchers at Northwestern University experimented with adding zirconium to a scandium-aluminum alloy. They discovered the unexpected
spontaneous formation of scandium-rich grains coated in zirconium-

HAFNIUM CAN REPLACE SILICON IN TRANSISTORS

According to Gordon Moore (1929–   ), cofounder of Intel, "the
biggest change in transistor technology since the late 1960s" is
the use of hafnium in transistors. Silicon has historically been the
semiconductor of choice in the transistors that regulate signal
switching in computer chips.

Semiconductors rely on the ability of electrons to jump
across gaps between accessible energy levels in the insulating
layer of a material in order for current to flow. Electrons can be
forced to jump between levels or *bands* if they are fed the right
amount of energy. An applied electric field or light absorption
can therefore cause current flow in semiconductors. It is this
property that makes semiconductor materials useful in circuits—small amounts of current can be made to flow at prescribed moments. This is an important property in transistors,
optoelectronics, collision avoidance radar, and solar cells.

For decades, silicon dioxide has been a highly successful material to use in the insulating layer. In order to advance, however,
computer and other electronic applications rely on minimizing
the size of the chips. Silicon dioxide does not seem to be able to
handle the narrowing gap: It tends to leak more charge as the gap
becomes thinner. More power must then be added to allow the
same current to flow—an undesirable effect, as it adds expense.

Hafnium dioxide, with its higher *dielectric constant,* is proving promising in this application, as it performs well at thick-
nesses as small as 40 nanometers without excessive charge
leakage. The switch to hafnium-based semiconductors is not
simple, however. There are problems with electrode compatibility that may make the transition costly in the short run. Research
and development in this field continues.

rich shells. Extremely tiny in size (*nanometer* scale), these grains seem to be the reason that Al-Sc-Zr alloys exhibit unusually high strength and thermal stability.

Experiments show that adding zirconium to an Al-Sc alloy suddenly and intensely disorders the material, so that scandium and zirconium become distributed fairly evenly before they begin to form grains (or *precipitants*). Scandium has a higher *diffusion coefficient* than zirconium, possibly due to atomic size or mass differences. This means that the scandium finds it easier to move around, collide, and form grains early on, while the zirconium moves more slowly, latching on to the grains after they have already formed. The resulting shell of zirconium surrounding a bead of scandium acts as a protective coating that keeps the grain from growing as a result of collisions with other particles.

As reported in the June 2006 issue of *Nature Materials,* researcher Peter Voorhees states that insights from this study will allow scientists to "design other alloys that form core-shell *nanoparticles* and, as a result, create new alloys with greatly enhanced properties."

TECHNOLOGY AND CURRENT USES OF THE TITANIUM GROUP

Beginning in the middle of the 20th century, titanium, zirconium, and hafnium became important elements in the world's economy. The main use of titanium is to strengthen alloys and to promote corrosion resistance in applications that include airframes, engines, and submarines. Titanium is also used in golf clubs, surgical instruments, and prosthetic devices.

Military uses of titanium include its use in cannons, guided missiles, and armor plate for tanks. Titanium tetrachloride is an important catalyst in the *polymerization* of ethylene. Liquid titanium tetrachloride is used in smokescreens and skywriting. Titanium dioxide is widely used as a white paint pigment; as a whitener in the production of paper, rubber and leather products, ink, and ceramics; and as a finishing agent in porcelain enamels. The Guggenheim Museum in Bilbao, Spain, designed by Canadian-American architect Frank Gehry (1929–), incorporates highly reflective external titanium panels as an aesthetic

Strong, corrosion-resistant titanium is a useful component in the manufacture of gears. *(Christian Lagerek/Shutterstock)*

feature. Barium titanate is used as a transducer in the interconversion of sound and electrical energy.

In the form of the mineral zircon, zirconium is an important diamond-like gemstone. Metallic zirconium is used to coat uranium fuel rods for use in nuclear reactors. Zirconium hydride is used in flares, detonators, and ceramics. Zirconium diboride is used to coat ball bearings. Zirconium oxide is used in glassmaking to make the glass resistant to attack by alkaline solutions. Zirconium oxide is also used to make laboratory *crucibles*. Recent research and development has given rise to a bulk metallic glass made from pure zirconium—the first time a glass has been fabricated from a single element. Pure zirconium glass remains stable at temperatures far higher than *composite* metallic glasses. This property makes it intriguing for use in medical products and electronics as well as sports equipment.

Various zirconium compounds are used in *polymers,* in treating textiles, in *catalysis,* in superconductors (as an alloy with niobium), and

The Bilbao Museum in Spain incorporates highly reflective external titanium panels. *(Cornel Achirei/Shutterstock)*

in ion-exchange columns. Zirconium compounds can also be found in deodorants and in antidotes for poison ivy.

The main use of hafnium is in the control rods of nuclear reactors; hafnium is used because it is a strong absorber of neutrons. Hafnium is also used in gas-filled incandescent light bulbs and in tungsten and molybdenum alloys, and is challenging silicon as the semiconductor of choice in computer chips. A recently enhanced ability to detect hafnium isotopes in zircon is informing geologists about how certain types of granite form and may help scientists understand the mechanisms of chemical differentiation in Earth's mantle and crust.

The ability to substitute titanium for lead in the paint and paper industries, as well as titanium's excellent strength in alloys, ensures its continued use in the years to come. As alternatives to fossil fuels are developed and the nuclear industry makes a comeback, demand for zirconium and hafnium is likely to continue.

THE ASTROPHYSICS OF THE VANADIUM GROUP:
V, Nb, Ta

The elements in the vanadium group can provide clues to astrophysicists who study processes in supernovae or the formation of Earth's Moon and the *rocky planets*.

THE BASICS OF VANADIUM

Symbol: V
Atomic number: 23
Atomic mass: 50.9415
Electronic configuration: [Ar]$3d^34s^2$

T_{melt} = 3,470°F (1,910°C)
T_{boil} = 6,165°F (3,407°C)

Abundance in Earth's crust = 34 ppm

Isotope	Z	N	Relative Abundance
$^{50}_{23}$V	23	27	0.25%
$^{51}_{23}$V	23	28	99.75%

Vanadium

2
8
11
2

V 23

+2 +3 +4 +5
50.9415
9.6X10^{-7}%

1910°
3407°

THE BASICS OF NIOBIUM

Symbol: Nb
Atomic number: 41
Atomic mass: 92.9064
Electronic configuration: [Kr]$4d^45s^1$

T_{melt} = 4,491°F (2,477°C)
T_{boil} = 8,571°F (4,744°C)

Abundance in Earth's crust = 17 ppm

Isotope	Z	N	Relative Abundance
$^{93}_{41}$Nb	41	52	100%

Niobium

2
8
18
12
1

Nb 41

+3 +5
92.90638
2.28X10^{-9}%

2477°
4744°

THE BASICS OF TANTALUM

Symbol: Ta
Atomic number: 73
Atomic mass: 180.9479
Electronic configuration:
[Xe]$4f^{14}5d^36s^2$

T_{melt} = 5,463°F (3,017°C)
T_{boil} = 9,856°F (5,458°C)

Abundance in Earth's crust = 1.7 ppm

	Tantalum	
2		3017°
8	**Ta**$_{73}$	5458°
18		
32		
11		
2		+5
		180.9479
		6.75X10^{-11}%

Isotope	Z	N	Relative Abundance
$^{180}_{73}$Ta	73	107	0.01%
$^{181}_{73}$Ta	73	108	99.99%

The brightness of a supernova depends strongly on the amount of vanadium 47 created in the explosion. As detailed in the previous section, "The Astrophysics of the Titanium Group," titanium 44 nuclei created in supernovae decay by emitting high-energy photons that can be detected by satellites built for that purpose. But collisions with alpha-particles can destroy Ti-44 before it can spontaneously decay, a process that results in the unstable isotope vanadium 47.

$$^{44}_{22}\text{Ti} + ^4_2\alpha \rightarrow ^{47}_{23}\text{V} + ^1_1\text{p}$$

In this reaction, no photon is produced, so the supernova is dimmer than if all the Ti-44 were to undergo radioactive decay. Recent experiments have shown that alpha collisions are more prevalent than expected, meaning that a supernova gamma-ray signal will be diminished by about 25 percent. This will need to be considered by designers of instruments that are intended to detect gamma rays from supernovae.

Observations of the elements vanadium 51 and tantalum 180 in supernovae spectra may prove important as tools for understanding *neutrino oscillations*. The neutrino process that helps produce these vanadium group *nuclides* occurs when the enormous flux of neutrinos formed in a supernova rushes through the shell of the exploding star, knocking protons or neutrons off heavy nuclei in their paths. Tantalum 180 that is made in this way may turn out to be a particularly sensitive indicator of *electron neutrino* temperature at the moment of the explosion. Particle physicists have only recently shown that neutrinos do have mass. Experiments have so far been sufficient to put limits on the possible mass range, but much more work needs to be done before the mass of each flavor of neutrino is known. Studies of tantalum and electron neutrino temperature in supernovae may provide important clues to aid in this research.

Another astronomical area of research where tantalum has been important is in determining the age of Earth's Moon. It is known that the Moon formed during the catastrophic collision of another planet-sized object with Earth, which would have liquefied any core that might have existed at that time. Studies of elements in Earth's core can, therefore, be correlated with the impact and subsequent cooling. The tungsten 182 isotope (a decay product of hafnium 182) is a key indicator of core formation because, once formed, it tends to concentrate in the core rather than in Earth's mantle. By this measure, the impact forming the Moon was estimated at 30 million years after the solar system began to *accrete*. This analysis, however, failed to take into account the formation of tungsten by the decay of ^{182}Ta. When both modes of radioactive decay are taken into account, the data indicate a lunar formation around 60 million years after the solar system began. There is a large margin of error, however, that needs to be narrowed before scientists can arrive at a reliable age for the Moon. Further exploration of the Moon and Mars could provide samples to help answer the question.

The timing of early and pre–solar system events and processes may also be answered by studies of extinct *radionuclides* like niobium 92, which is created in supernovae and decays to ^{92}Zr with a half-life of 36 million years. Any amount of this isotope that still remained during the solar system's accretion can be measured by detection of its Zr-92 daughter, the abundance of which is higher in some meteorites than in others. This sort of selective formation is of great interest to planetary geologists. More study in this area could lead to a better understanding of why the planets vary in elemental composition.

DISCOVERY AND NAMING OF VANADIUM, NIOBIUM, AND TANTALUM

In 1801, the Mexican *mineralogist* Andrés Manuel del Río (1764–1849) found a new metal in an ore from Hidalgo. The metal's properties were very similar to those of chromium. At first, he named the new element *panchromium* because its compounds represented an entire spectrum of colors. Later, del Río changed the name to *erythronium* from the reddish color of its alkaline salts. He later decided that the metal's properties seemed, in fact, to be so identical to chromium that perhaps he had made a mistake and that the metal was chromium after all.

In 1831, the Swedish physician and chemist Nils Gabriel Sefström (1787–1845) found what he believed to be a new element in an ore from a mine in Småland. Analysis convinced him that the element was del Rio's erythronium and was not simply a sample of chromium. The name *vanadium* came from the Norse goddess of fertility Vanadis, who was known for her beauty, in reference to vanadium's very colorful compounds.

After Sefström's death, the German chemist Friedrich Wöhler (1800–82) demonstrated that del Río's original ore did in fact contain vanadium. Even then, however, pure vanadium had not been isolated. In 1867, Henry Roscoe (1833–1915) succeeded in isolating pure metallic vanadium by reducing vanadium chloride (VCl_3) with hydrogen gas (H_2).

The English chemist Charles Hatchett (1765–1847) spent his career performing research in analytical and mineralogical chemistry. In 1801, Hatchett became interested in a heavy black ore called *columbite* that was kept in the British Museum and reportedly had been sent from New England during American colonial times. His analysis led him to believe that the ore contained an unknown metal that he named *columbium* from the name of the ore and to which he assigned the symbol Cb.

Hatchett was unable to produce pure columbium. It was not until the 1860s that a sample of the pure metal was finally obtained and not until the early 1900s that the atomic mass was determined. Later in the 1900s, the name of the element was changed to *niobium* (symbol Nb) to reflect its association in ores with tantalum. In Greek mythology,

Tantalus was the father of Niobe, whom mythology said was turned to stone while mourning the loss of her children.

Tantalum was discovered by the Swedish chemist and mineralogist Anders Gustaf Ekeberg (1767–1813), a professor at the University of Uppsala in the Swedish city of Uppsala. In 1802, Ekeberg analyzed samples of the minerals gadolinite and yttrotantalite and concluded that both minerals contained an unknown metal. He named the new element *tantalum* because it was so tantalizing to try to isolate it. The origin of the word *tantalize* itself is derived from Greek mythology. According to legend, Tantalus was a king who committed such horrific crimes that he was condemned to stand in water in Hades for all eternity. Any time he tried to drink the water, the water receded below the level at which he could reach it. Any time Tantalus tried to reach the fruit that was hanging above him, the fruit receded to a height at which he could not reach it.

Because tantalum and niobium tend to occur together and have almost identical properties, at first some chemists thought that they were the same element. It was not until 1846 that the German analytical chemist Heinrich Rose (1795–1864) proved that they are different elements. As described previously, niobium was originally called *columbium* and continued to be called columbium in the United States well into the 20th century. Rose, however, gave element 41 the name *niobium*. That name was adopted in Europe during the 19th century. In 1951, the International Union of Pure and Applied Chemistry (IUPAC) tried to resolve the conflict between the two names by officially declaring the name to be *niobium*. Nevertheless, only later was the name *niobium* finally adopted in the United States.

THE CHEMISTRY OF THE VANADIUM GROUP

Vanadium is element 23, with a density of 5.8 g/cm^3. It is the 19th most abundant element in Earth's crust. A number of ores contain vanadium. These include vanadinite ($Pb_5[VO_4]_3Cl$), a bright red or orange-red crystal often found mixed with lead ores and commonly sold in rock shops; dechenite ($PbZn[VO_3]_2$); descloizite ($PbZn[VO_4][OH]$), a translucent orange-red to reddish brown mineral with a greasy *luster;* pucherite ($BiVO_4$); roscoelite (mica with V_2O_3); and volborthite ($Cu_3V_2O_7[OH]_2 \cdot 2H_2O$), a green, yellow, or brown mineral. In addition, the following are two vanadium minerals that are radioactive because they also contain uranium:

- tyuyamunite $(Ca[UO_2]_2V_2O_8 \cdot nH_2O)$, a greenish-yellow to yellow mineral
- carnotite $(K_2[UO_2]_2V_2O_8 \cdot 3H_2O)$, a bright yellow mineral that is also an important source of uranium.

Vanadium can also be recovered from magnetite (Fe_3O_4) and bauxite (Al_2O_3).

In the United States, vanadinite is found principally in Arizona and New Mexico. Vanadium is also imported from countries like Canada, China, and South Africa. Vanadium is extracted from its ores by dissolving the ore in concentrated hydrochloric acid (HCl), adding excess ammonium chloride (NH_4Cl), and then evaporating the solution. Upon

A broken spillway from a vanadium mine in northwest China in 2008 contaminated two nearby rivers with ore tailings. *(Reuters Limited)*

heating, ammonium vanadate converts to vanadium pentoxide (V_2O_5), from which the metal can be obtained.

Niobium is element 41, with a density of 8.55 g/cm³. Niobium is the 33rd most abundant element in Earth's crust. The principal mineral containing niobium is columbite ($[Fe, Mn][Nb, Ta]_2O_6$), which is gray to brownish black and which also contains tantalum. Columbite is found in association with granite. Another source of niobium is samarskite, a black or brownish mineral found in granite or heavy sands as a mixture of yttrium, cerium, uranium, iron, niobium, tantalum, and titanium oxide. The production of pure niobium is very difficult and usually gives an unsatisfactory mixture.

Tantalum is element 73, with a relatively high density of 16.6 g/cm³. Tantalum ranks 51st in order of abundance of elements in Earth's crust and is mainly obtained as a by-product of tin processing. The principal mineral is tantalite ($FeTa_2O_6$), which is a black to red-brown or colorless mineral found in association with vanadium and niobium ores. As mentioned previously, columbite is also a source of tantalum. Most tantalum used in the United States is imported from countries like Australia, Canada, and China. In addition, because of its scarcity, a significant amount of tantalum is recycled.

All three metals are gray or silvery white in color. They are very hard metals, yet they are also malleable and ductile. The vanadium group metals are *semi-noble,* meaning that they are resistant to oxidation. As such, they tend to withstand tarnishing because they react with atmospheric oxygen only at high temperatures. Their resistance to oxidation is also demonstrated by their tendency not to be attacked by salt water. Vanadium and niobium can be dissolved in nitric acid, but tantalum is unaffected. To dissolve tantalum requires the more powerful oxidizing agent *aqua regia,* a mixture of concentrated hydrochloric and nitric acids.

Vanadium is the first metal in the periodic table to exhibit chemistry in a large number of oxidation states—in this case, "+2," "+3," "+4," and "+5". In addition, vanadium is the first element in the first row of transition metals to form a variety of very colorful compounds and complex ions. (The colors are due to the way light interacts with the 3d electrons in first-row transition elements.) The colors associated with each oxidation state are distinctive enough that they can be used to identify what oxidation state vanadium is in.

Vanadium occurs as the vanadous ion in the "+2" oxidation state. The vanadous ion forms vanadous oxide (VO), which is sufficiently metallic-looking that it is sometimes mistaken for the metal itself. Vanadous sulfate (VSO_4) has a deep violet color, as do other V^{2+} salts.

Vanadium in the "+3" state occurs as the vanadic ion. Its chemistry is similar to that of the ferric ion (Fe^{3+}), with the exception that Fe^{3+} is very difficult to oxidize to higher oxidation states, whereas V^{3+} can be oxidized to the "+4" and "+5" states. The salts, hydroxide, and oxide of V^{3+} tend to be green. Examples include $V(OH)_3$, V_2O_3, VCl_3, $VOCl$, $V_2(SO_4)_3$, VN, and V_2S_3. The vanadic ion also forms many *complex salts,* including $(NH_4)_3VF_6$, $K_3V(CN)_6$, and $K_3V(CNS)_6$.

In the "+4" state, vanadium forms vanadium dioxide (VO_2). More commonly, the "+4" oxidation state is represented by vanadites, where vanadite refers to the $V_4O_9^{2-}$ ion. An example is $K_2V_4O_9 \cdot 7H_2O$. Vanadites tend to have a dark black color.

As a *group VA* element, the highest oxidation state to be expected of vanadium is "+5." In the "+5" state, the most important compound of vanadium is the oxide, V_2O_5. In addition, vanadium in the "+5" state occurs as the *metavanadate* ion (VO_3^-) or as the *pervanadyl* ion (VO_2^+). The pervanadyl ion forms when V_2O_5 dissolves in acid, as shown in the following chemical equation:

$$V_2O_5 \text{ (s)} + 2 \text{ H}^+ \text{ (aq)} \rightarrow 2 \text{ VO}_2^+ \text{ (aq)} + H_2O \text{ (l)}.$$

The pervanadyl ion has a tendency to polymerize, as shown in the following equation:

$$10 \text{ VO}_2^+ \text{ (aq)} + 8 \text{ H}_2O \text{ (l)} \rightarrow H_2V_{10}O_{28}^{4-} \text{ (aq)} + 14 \text{ H}^+ \text{ (aq)}.$$

Metavanadic acid (HVO_3) is a yellow solid. A number of salts of metavandic acid can be formed, including sodium metavanadate ($NaVO_3$), potassium metavanadate (KVO_3), ammonium metavanadate (NH_4VO_3), and barium metavandate ($Ba[VO_3]_2$). In addition, there are a variety of sodium salts that include sodium orthovanadate (Na_3VO_4), sodium pyrovanadate ($Na_4V_2O_7$), and sodium hexavanadate

($Na_2H_2V_6O_{17}$). Ammonium sulfide [$(NH_4)_2S$] reacts with vanadates to form the thiovanadate ion (VS_4^{4+}), where the prefix *thio* in chemical nomenclature refers to substances in which sulfur atoms have replaced oxygen atoms. Adding acid to the solution precipitates vanadium (V) sulfide (V_2S_5). The distinctive violet-red color of ($NH_4)_3VS_4$ is used to detect the presence of vanadium in solution.

The chemical and physical properties of niobium and tantalum are almost identical. Niobium tends to be a very *passive (noble)* metal. Niobium's chemistry is limited primarily to the "+3" and "+5" oxidation states. The "+3" ion—called the niobic ion (Nb^{3+})—imparts a blue color to solutions. However, the "+3" ion is very unstable, resulting in rapid oxidation by atmospheric oxygen to the "+5" oxidation state.

There are a number of niobium compounds in which niobium is in the "+5" oxidation state, including niobium pentoxide (Nb_2O_5), niobium pentachloride ($NbCl_5$), and niobium nitride (Nb_3N_5).

In terms of stable compounds, tantalum's chemistry is limited to the "+5" oxidation state. When tantalum burns in air, tantalum pentoxide is formed, as shown by the following chemical equation:

$$4\ Ta\ (s) + 5\ O_2\ (g) \rightarrow 2\ Ta_2O_5\ (s).$$

Tantalum readily forms *pentahalides,* including TaF_5, $TaCl_5$, and $TaBr_5$. When $TaCl_5$ is heated in ammonia, the red-colored nitride, Ta_3N_5, is formed.

REDUCING THE SIZE OF PARTICLE ACCELERATORS

High-energy particle accelerators are the tools of choice for physicists who want to understand matter at the subatomic level. Research scientists use electric and magnetic fields to accelerate beams of charged particles like protons and electrons to extremely high velocities. These particles are then aimed either at a chosen material like hydrogen gas or made to collide with another beam of particles. Because the beams have such high energy, upon impact they are able to break apart very strong bonds in the target particles, which may be other subatomic particles or atomic nuclei. Detection of the resulting collision fragments often results in the discovery of new particles and processes. The world's largest particle accelerator, the Large Hadron Collider

TRINIOBIUM TIN IN SUPERCONDUCTING MAGNETS

High magnetic fields are important in particle accelerators, fusion reactors, and magnetic resonance imaging, among other scientific applications. The strongest magnetic fields can be produced by winding conducting wire around a cylinder and then passing a current through the wire, a phenomenon discovered by Hans Christian Ørsted (1777–1851) in 1820. The magnetic field at the center of the cylinder becomes stronger as the current is increased. Unfortunately, most materials don't stand up against very high currents and field intensities.

The simple compound triniobium tin (Nb_3Sn), however, which is superconducting at temperatures below 18 K, has shown the ability to withstand magnetic fields as high as 30 tesla—a field nearly three times that which can be tolerated by niobium-titanium, its closest competitor. Because Nb_3Sn is a

(continues)

Workers at Brookhaven National Laboratory in New York install a section of superconducting triniobium-tin magnet coil. *(Brookhaven National Laboratory)*

(continued)

superconductor, current flow experiences virtually no resistance as it travels through the wire, so no power is lost as heat. However, fabricating wire from this extremely brittle material is difficult and expensive. The need for extremely high magnetic fields in high-energy particle accelerators is largely driving research into new fabrication methods. The ability to produce such high-sustained magnetic fields would allow physicists to accelerate electrons and positrons to speeds approaching the velocity of light, opening up the possibility to explore new states of matter and conditions similar to those that existed only moments after the big bang.

(LHC) near Geneva, Switzerland, for example, will simulate conditions very early in the evolution of the universe—just moments after the big bang—which may help with an understanding of how pure energy can become mass.

In order for protons to reach high enough energies to initiate such exotic states of matter, they need to be accelerated over great distances by strong magnetic fields—up to 14 trillion electron volts in the case of the Large Hadron Collider, a circular accelerator 17 miles (27 km) in circumference. The giant magnet coils that produce the acceleration are made of superconducting niobium-titanium cables clad with copper. The Nb-Ti metal compound can withstand high currents and magnetic fields up to 10 tesla (T), but a newly developed material, trinobium tin (Nb_3Sn), is proving even hardier, performing well in fields as high as 30 T. This is important because a stronger field will accelerate particles at a higher rate, allowing for smaller accelerators in the future.

TECHNOLOGY AND CURRENT USES OF THE VANADIUM GROUP

Most vanadium and niobium is used in various alloys. Tantalum is an expensive metal, so there are fewer economical applications.

More than half the vanadium produced in the United States is alloyed with steel to increase strength and toughness. V_2O_5 and NH_4VO_3 are important catalysts in the chemical industry and are used in the manufacture of nylon and the production of sulfuric acid, acetaldehyde, and oxalic acid. V_2O_5 also produces a black pigment for the dye industry.

Vanadium steel is so strong that it is used to make armor plates for military vehicles. It is also used in engine parts and to make the frames of high-rise buildings. Vanadium is used in nuclear reactors to reduce corrosion. Vanadium compounds are used in the production of ceramics. Alloyed with gallium, vanadium can be used to make superconducting magnets.

Niobium is used in stainless steel, high-temperature alloys and in superconducting alloys. Niobium is also used in nuclear reactors. Compounds of niobium and titanium are used in the windings of the superconducting magnets that are needed in high-energy particle accelerators, and triniobium tin may allow for even higher magnetic fields to accelerate particles to speeds near the speed of light.

The first application of tantalum occurred in the early 20th century, when it was used to make wire. Most tantalum is used to make capacitors in applications that include smoke detectors, heart pacemakers, and automobiles. In addition, tantalum has dental and surgical uses and forms alloys with a variety of other metals.

3

The Chromium and Manganese Groups

Chromium and manganese and the other elements in their families are important to society and to modern industry. In this chapter, readers will be introduced to the stories of their discoveries, chemical and physical properties, and important uses.

The chromium group, labeled VIA in the periodic table, consists of chromium (Cr, element 24), molybdenum (Mo, element 42), and tungsten (W, element 74). Chromium and tungsten are probably the most familiar metals of the three, given chromium's role in the manufacture of stainless steel and tungsten's use as the filament in incandescent lightbulbs. Compounds of these elements exist in a variety of oxidation states. Molybdenum and tungsten can exist in the "+2," "+3," "+4," "+5," and "+6" states, and chromium most commonly occurs in the "+2," "+3," and "+6" states.

THE BASICS OF CHROMIUM

Symbol: Cr
Atomic number: 24
Atomic mass: 51.996
Electronic configuration: $[Ar]4s^1 3d^5$

T_{melt} = 3,465°F (1,907°C)
T_{boil} = 4,840°F (2,671°C)

Abundance in Earth's crust = 140 ppm

	Chromium	
2		1907°
8	**Cr**₂₄	2671°
13		
1		
	+2 +3 +6	
	51.9961	
	0.000044%	

Isotope	Z	N	Relative Abundance
$^{50}_{24}$Cr	24	26	4.35%
$^{52}_{24}$Cr	24	28	83.79%
$^{53}_{24}$Cr	24	29	9.50%
$^{54}_{24}$Cr	24	30	2.36%

THE BASICS OF MOLYBDENUM

Symbol: Mo
Atomic number: 42
Atomic mass: 95.94
Electronic configuration: $[Kr]5s^1 4d^5$

T_{melt} = 4,753°F (2,623°C)
T_{boil} = 8,382°F (4,639°C)

Abundance in Earth's crust = 1.1 ppm

	Molybdenum	
2		2623°
8	**Mo**₄₂	4639°
18		
13		
1	+6	
	95.94	
	8.3X10⁻⁹%	

Isotope	Z	N	Relative Abundance
$^{92}_{42}$Mo	42	50	14.84%
$^{94}_{42}$Mo	42	52	9.25%
$^{95}_{42}$Mo	42	53	15.92%
$^{96}_{42}$Mo	42	54	16.68%
$^{97}_{42}$Mo	42	55	9.55%
$^{98}_{42}$Mo	42	56	24.13%
$^{100}_{42}$Mo	42	58	9.63%

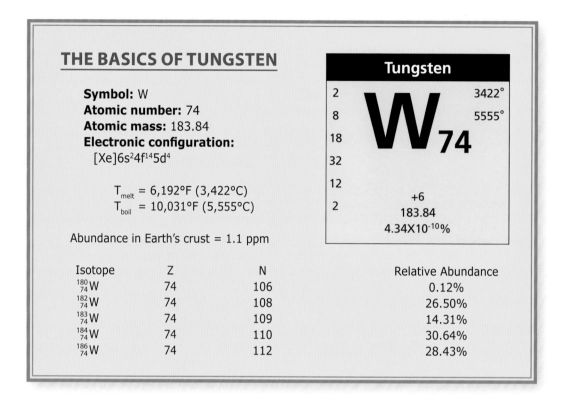

THE BASICS OF TUNGSTEN

Symbol: W
Atomic number: 74
Atomic mass: 183.84
Electronic configuration:
  [Xe]$6s^2 4f^{14} 5d^4$

T_{melt} = 6,192°F (3,422°C)
T_{boil} = 10,031°F (5,555°C)

Abundance in Earth's crust = 1.1 ppm

Tungsten

2		3422°
8		5555°
18	**W** 74	
32		
12		+6
2		183.84
		4.34X10^{-10}%

Isotope	Z	N	Relative Abundance
$^{180}_{74}$W	74	106	0.12%
$^{182}_{74}$W	74	108	26.50%
$^{183}_{74}$W	74	109	14.31%
$^{184}_{74}$W	74	110	30.64%
$^{186}_{74}$W	74	112	28.43%

The manganese metals are labeled Group VIIA and consist of manganese (Mn, element 25), technetium (Tc, element 43), and rhenium (Re, element 75). Manganese is a familiar element that is essential to living organisms and is a component of steel alloys. Technetium is unusual, as it is the lightest element that does not have any stable isotopes. Although two of technetium's isotopes have half-lives on the order of a million years, Earth is billions of years old. Thus, all of the technetium that was present when Earth formed has decayed away. Nevertheless, substantial quantities of technetium are produced in nuclear reactors and used daily in medical diagnosis and treatment. Prior to 1994, it was believed that there were no rhenium minerals. In that year, however, an ore of rhenium sulfide was discovered in Russia.

THE ASTROPHYSICS OF THE CHROMIUM GROUP: CR, MO, W

While members of the chromium group are similar in many ways, they are synthesized very differently in stars. Chromium is an astrophysical alpha-process element like oxygen, neon, argon, and others. Alpha-process elements are synthesized by sequential absorption of alpha particles, which have four nuclei, so they all have mass numbers that are multiples of 4. Chromium 48 generally forms by capture of an alpha (helium nucleus) onto titanium 44.

$$^{44}_{22}\text{Ti} + ^{4}_{2}\alpha \rightarrow ^{48}_{24}\text{Cr}$$

This process occurs during the *main sequence* lifetime as a consequence of interactions in the star's core, as well as during supernovae explosions, which distribute chromium into the interstellar medium. An odd feature of chromium abundance in the Sun is that the Cr^+ ion seems to be more prevalent in the solar atmosphere than neutral chromium. Further work on solar models that include nonequilibrium (i.e., *convection*) effects may clear up this "chromium ionization imbalance problem."

Molybdenum and tungsten, both being heavier than iron, cannot be formed by fusion in stellar cores. Molybdenum 100 is synthesized in supernovae via the rapid capture by iron nuclei of a succession of neutrons, which is called the r-process. Molybdenum 96, however, builds up over thousands of years in the atmospheres of large-mass stars via neutron capture. Because this synthesis proceeds relatively slowly due to a low density of neutrons, it is called the s-process. The ^{95}Mo, ^{97}Mo, and ^{98}Mo isotopes form by a series of r- and s-process events. In general, molybdenum is considered to be an s-process element because at least 50 percent of solar molybdenum originated in this way.

Tungsten (W) also eventually forms by the slow buildup of neutrons onto seed nuclei. Examinations of the relative abundance of ^{182}W and ^{184}W over ^{183}W in silicon carbide *stardust* by scientists from the Australian National University indicate that *asymptotic giant branch* (AGB) stars are a dominant source of tungsten. An unexpected *enrichment* in ^{186}W has recently helped researchers come to a new understanding about the activation of tungsten formation during thermal pulses in AGB stars.

DISCOVERY AND NAMING OF CHROMIUM, MOLYBDENUM, AND TUNGSTEN

Although chromium is the most familiar element in Group VIA, it was the last of the naturally occurring elements in this group to be discovered. Tungsten was the first to be discovered. The presence of tungsten in the minerals wolframite and scheelite was recognized in the 1700s, but at that time it was not isolated or identified as a new element. *Wolframite,* or *wolfram,* comes from the German and translates as "wolf froth," presumably a description of the mineral's appearance. It remained for two Spanish brothers to recognize tungsten as a new element. Their names were Don Juan José de Elhuyar y de Zubice (1754–96) and Don Fausto de Elhuyar (1755–1833).

The brothers visited Carl Wilhelm Scheele (1742–86), a prominent Swedish chemist for whom the mineral scheelite (calcium tungstate, $CaWO_4$) was later named. Scheele described to them his discovery of tungstic acid and suggested that tungstic acid might possibly be reduced to a new metallic element. Subsequent to their return home, in 1783, the de Elhuyars began their own investigations and succeeded in obtaining globules of tungsten by heating tungstic acid and charcoal in a crucible. The two brothers gave the new element the name *wolfram* after the mineral from which they obtained it. The name *wolfram* continues to be used to this day in Germany, and is the source of the symbol W for the element. The name *tungsten* was also in use and had been adopted in countries like France and England. (The word *tungsten* means *heavy stone.*) Although the IUPAC settled the dispute by adopting tungsten as the official name, the element has retained the symbol W.

The de Elhuyar brothers became quite famous in Spain. In 1786, King Charles of Spain sent Don Juan José to Columbia in South America to be the director of the Spanish mines in that country. Don Fausto was sent to Mexico to be the director of mines there.

The mineral molybdenite (MoS_2) was known in the mid-1700s. Its appearance is so similar to graphite that initially it was confused with graphite. In 1778, Scheele studied molybdenite and isolated a white solid he called molybdic acid. Although it was suggested to him that molybdic acid probably contained a new metal, his equipment was insufficient to pursue the idea. Instead, Scheele suggested to the Swedish chemist Peter Jacob Hjelm (1746–1813) that Hjelm should attempt the reduction of molybdic acid. In 1781, Hjelm heated molybdic acid with carbon in a crucible and succeeded in obtaining the metal molybdenum. The name *molybdenum* was derived from a Greek word that indicated molybdenum's association with lead ore.

The discovery of chromium is attributed to the French chemist Nicolas-Louis Vauquelin (1763–1829). The son of a poor farmer, Vauquelin grew up in extreme poverty. He did so well in school, however, that he was apprenticed to an *apothecary*. Later, he went to Paris, where he became the assistant to the French chemist Antoine-François de Fourcroy (1755–1809). Vauquelin eventually became an inspector of mines and professor at the School of Mines. A red ore from Siberia (called crocoite or crocoisite) had been studied for several years without a successful analysis. In 1797–98, Vauquelin began his own investigations of crocoite. Vauquelin pulverized his sample and obtained the element lead plus salts of an unknown element. In 1798, he isolated what would later be identified as chromium trioxide and reduced it to chromium metal by heating it with charcoal. When it was discovered that the new element had many brightly colored compounds, Fourcroy recommended the name *chromium* for the new element after the Greek word *chroma,* which means *color.*

Also in 1798, Vauquelin analyzed a green emerald and discovered that the green color was due to the presence of chromium. Later investigators discovered that the red color of rubies is also due to chromium. In both cases, chromium is present in very tiny quantities.

THE GEOLOGY OF THE CHROMIUM GROUP

Chromium has a density of 7.2 g/cm^3 and ranks 21st in abundance on Earth. It is found in a number of ores, including chromite ($FeCr_2O_4$), an igneous black to brownish-black mineral, and crocoite (lead chromate, $PbCrO_4$), an orange-red or yellow mineral. Today, most of the chromium used in the United States is imported from South Africa, Turkey, Zimbabwe, and Yugoslavia. In 1817, chromium was found in a meteorite that landed in Siberia. Chromium as the chromic ion Cr^{3+} is an essential element in animals. The chromic ion is necessary for the *metabolism* of *glucose.*

Molybdenum has a density of 10.2 g/cm^3 and ranks 54th in abundance on Earth. Molybdenum is found in a limited number of ores, including molybdenite (MoS_2), a very soft, gray mineral formed in

Chromium can be polished to a high luster. *(Theodore Gray/ Visuals Unlimited)*

hydrothermal vents, and wulfenite ($PbMoO_4$), a fairly common mineral that ranges from orange to yellow, or brown to greenish brown in color. In appearance, molybdenite resembles mica. Some of the world's largest molybdenum mines have been in the United States in the states of Colorado and Nevada. Molybdenum is also obtained as a commercially important by-product of copper mining in the western states of Arizona, New Mexico, Montana, and Utah. Significant molybdenum mines are found around the world, including mining operations in countries like Canada, China, Chile, Mexico, and Russia.

Tungsten has a very high density of 19.3 g/cm^3 (the same density as gold). It ranks 58th in abundance. Tungsten is mainly found in ores as an oxide. Examples include tungstite ($WO_3 \cdot H_2O$), a very soft, yellow mineral; wolframite ($FeMn[WO_4]_2$), a brownish-black mineral associated with granite; and scheelite ($CaWO_4$), a white, colorless, gray, yellow, orange-yellow, brownish green, reddish, or purple mineral that is the most important source of tungsten.

THE CHEMISTRY OF THE CHROMIUM GROUP

The maximum oxidation state of elements in Group VIA is "+6." In this respect, chromium, molybdenum, and tungsten share chemical properties similar to sulfur, which is in Group VIB. All three Group VIA elements exhibit several oxidation states, but the "+2" and "+3" states are common only with chromium. In that respect, chromium tends to more closely resemble vanadium and manganese than it does molybdenum and tungsten. Compounds in the "+6" state are common to all three elements, with most of molybdenum's and tungsten's chemistry being in that state.

Chromium's most stable oxidation state is the "+3," in which it exists as the violet chromic ion (Cr^{3+}). Thus, chromium in lower oxidation states—as the chromous ion (Cr^{2+})—is a good reducing agent (because it can be oxidized), and chromium in the "+6" state—as chromate (CrO_4^{2-}) or dichromate ($Cr_2O_7^{2-}$)—is a good oxidizing agent because it can be reduced. The Cr^{2+} ion is blue in aqueous solution, CrO_4^{2-} is yellow, and $Cr_2O_7^{2-}$ is orange. Both the chromous and chromic ions form a large number of compounds with bright colors, as shown in the following table:

	Cl^-	OH^-	NO_3^-	SO_4^{2-}	PO_4^{3-}	S^{2-}
Cr^{2+}	$CrCl_2$ (blue)	$Cr(OH)_2$ (br-yellow)	none	$CrSO_4$ (blue)	none	none
Cr^{3+}	$CrCl_3$ (violet)	$Cr(OH)_3$ (green)	$Cr(NO_3)_3$ (purple)	$Cr_2(SO_4)_3$ (violet)	$CrPO_4$ (blue-green)	Cr_2S_3 (brown)

Another important chromium (III) ion is chromite ($Cr_2O_4^{2-}$), which is green in color. Ferrous chromite ($FeCr_2O_4$) is used to line furnaces in the steelmaking industry.

The acid-base properties of chromium oxides vary with the oxidation state of the chromium. Chromous oxide (CrO) is basic because it is dissolved by acids but not by bases. Chromic oxide (Cr_2O_3) is *amphoteric* because it is dissolved by both acids and bases. Chromic trioxide (or chromium anhydride, CrO_3) is acidic; it is dissolved by bases but not by acids. The amphoteric nature of Cr_2O_3 is illustrated by the following reactions:

$$Cr_2O_3 \text{ (s)} + 6 \text{ HCl (aq)} \rightarrow 2 \text{ CrCl}_3 \text{ (aq)} + 3 \text{ H}_2\text{O } (l);$$

$$Cr_2O_3 \text{ (s)} + 2 \text{ NaOH (aq)} + 3 \text{ H}_2\text{O } (l) \rightarrow$$
$$2 \text{ Na}^+ \text{ (aq)} + 2 \text{ Cr(OH)}_4^- \text{ (aq)}.$$

In the first reaction, Cr_2O_3 is being dissolved by acid; in the second reaction, it is being dissolved by base.

In the "+6" oxidation state, chromium's two important ions are CrO_4^{2-} and $Cr_2O_7^{2-}$. Chromate behaves as a base in aqueous solution, and dichromate behaves as an acid. By adding acid to a chromate solution, chromate can be converted to dichromate, as shown by the following equation:

$$2\ CrO_4^{2-}\ (aq) + 2\ H^+\ (aq) \rightarrow Cr_2O_7^{2-}\ (aq) + H_2O\ (l).$$

In acidic solution, the dichromate ion is an exceptionally powerful oxidizing agent. Potassium dichromate ($K_2Cr_2O_7$) is often used in *quantitative analytical* chemistry. For example, the amount of iron in a compound or mixture can be determined by *titrating* the iron sample with $K_2Cr_2O_7$, as shown in the following reaction:

$$6\ Fe^{2+}\ (aq) + Cr_2O_7^{2-}\ (aq) + 14\ H^+\ (aq) \rightarrow$$
$$6\ Fe^{3+}\ (aq) + 2\ Cr^{3+}\ (aq) + 7\ H_2O\ (l).$$

If the iron was not initially in the form of ferrous ions (Fe^{2+}), it can be converted to ferrous ions before performing the titration.

Unfortunately, there are health hazards associated with chromate ions, dichromate ions, and other forms of chromium in its "+6" oxidation state, usually referenced as *hexavalent* chromium or Cr (VI). The National Institute for Occupational Safety and Health considers Cr (VI) to be a potential occupational carcinogen associated with an increased risk of lung cancer. Other possible adverse health effects associated with Cr (VI) exposure include skin problems; asthma; nasal irritation, rhinitis, nosebleed, and other respiratory problems; nasal and sinus cancer; eye irritation; and kidney and liver damage. Hexavalent chromium compounds are most safely handled by avoiding contact with the skin or inhalation of dust particles.

In *qualitative analytical* procedures, the presence of chromous or chromic ions can be detected by oxidizing them to the chromate ion in basic solution. Frequently, the oxidizing agent used is 3 percent hydrogen peroxide (H_2O_2), as shown in the following equation for the oxidation of $Cr(OH)_4^-$:

$$2\ Cr(OH)_4^-\ (aq) + 3\ H_2O_2\ (aq) + 2\ OH^-\ (aq) \rightarrow$$
$$2\ CrO_4^{2-}\ (aq) + 8\ H_2O\ (l).$$

Visual evidence that the reaction has occurred is the development of the yellow color associated with the chromate ion. The 3 percent H_2O_2 is the same strength of hydrogen peroxide that is sold at pharmacies.

If other species that might be present exhibit similar yellow colors, or colors that would mask chromate's color, additional tests may be performed. For example, a solution containing Ba^{2+} can be added. If CrO_4^{2-} is present, $BaCrO_4$ (the only yellow barium compound) will precipitate, as shown in the following equation:

$$Ba^{2+} (aq) + CrO_4^{2-} (aq) \rightarrow BaCrO_4 (s).$$

Another possibility is to add additional hydrogen peroxide in a solution of nitric acid. The acid converts chromate to dichromate, which, in turn, is oxidized by H_2O_2 to a chromium peroxide (CrO_5), as shown in the following equation:

$$Cr_2O_7^{2-} (aq) + 4\ H_2O_2 (aq) + 2\ H^+ (aq) \rightarrow 2\ CrO_5 (aq) + 5\ H_2O\ (l).$$

CrO_5 is very unstable, but it exhibits a characteristic blue color that cannot be mistaken for any other species that might also be present. Because the blue color quickly fades, liquid ether is sometimes added before the hydrogen peroxide. CrO_5 is more stable in ether, so shaking the solution extracts CrO_5 into the ether, where the blue color persists for a longer period of time.

Molybdenum and tungsten form compounds in the "+2," "+3," "+4," "+5," and "+6" oxidation states. In the "+2" state, both elements form chlorides ($MoCl_2$ and WCl_2) that are unstable in aqueous solutions. In the "+3" state, molybdenum exists in solution as an olive-green species. It forms a hydroxide ($Mo[OH]_3$), an oxide (Mo_2O_3), a sulfide (Mo_2S_3), and a phosphate ($Mo_3[PO_4]_2$). Tungsten does not form any simple compounds in the "+3" state.

The principal ore of molybdenum is molybdenite, in which molybdenum is present in the "+4" state as molybdenum sulfide (MoS_2). The "+4" ion itself is unstable, but stable *cyanide* complexes such as $K_4Mo(CN)_8$ may be prepared. Tungsten's compounds in the "+4" state include tungsten dioxide (WO_2) and tungsten tetrachloride (WCl_4).

(continued on page 52)

CHROMIUM AND THE MAGNETISM OF STAINLESS STEEL

Chromium and molybdenum are often added to stainless steel alloys to improve corrosion resistance. Although all stainless steel is predominantly iron, which is easily magnetized, the arrangement of atoms in the alloy dictates whether the particular steel in question will exhibit magnetic properties. The inclusion of Cr or Mo changes the fundamental structure of the alloy so that it is likely to incorporate body-centered cubic (BCC) crystals, rather than the face-centered cubics (FCC) that take shape when the alloy includes nickel, manganese, carbon, or nitrogen. Compound alloys

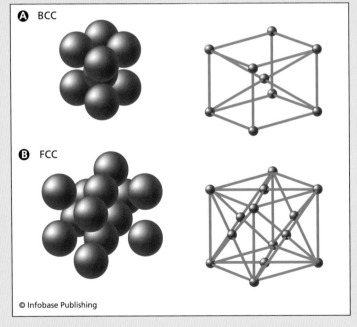

Ⓐ BCC

Ⓑ FCC

The inclusion of Cr or Mo can change the fundamental structure of a steel alloy so that it is likely to incorporate body-centered cubic (BCC) crystals, rather than the face-centered cubics (FCC) that take shape when the alloy includes nickel, manganese, carbon, or nitrogen.

Direction of External Applied Magnetic Field

© Infobase Publishing

In a ferromagnetic material, tiny magnetic domains typically point in random directions (top), and an external magnetic field can align them in the same direction (bottom).

like stainless steel *Type 304* contain chromium as well as nickel and manganese, so that both BCC and FCC regions are formed.

Materials undergo magnetization when tiny *magnetic domains* within the material become aligned in the same direction. (See chapter 6.) This occurs because of the influence of an external magnetic field that may be intentionally or unintentionally applied. The freedom of motion of these magnetic domains depends upon their positions within the metal lattice. While domains easily line up in BCC crystals, the FCC configuration seems to curtail their ability to rotate, so that even in a fairly strong external field, the steel does not become noticeably magnetic.

Thus, a side effect of adding Cr or Mo to stainless steel is that the material becomes *ferromagnetic*. Auto mechanics and construction workers find magnetized screwdrivers, for example, to be advantageous. Once magnetized, the magnetic field in such materials is typically preserved unless they suffer a mechanical jolt (like being dropped) or experience high temperatures.

(continued from page 49)

Tungsten (IV) compounds, however, tend to convert easily to the "+6" state.

In the "+5" state, molybdenum pentachloride ($MoCl_5$) is the only compound that does not contain oxygen. Species that contain oxygen include molybdenum hydroxide ($MoO[OH]_3$), molybdenum pentoxide (Mo_2O_5), and a deep-blue solid called "molybdenum blue" ($[MoO]_3[MoO_4]_2$). Tungsten exists in the "+5" state as the WO^{3+} or WO_2^+ ions, both of which are light green in color. Tungsten (V) also forms a pentachloride (WCl_5). In analogy to molybdenum blue, tungsten forms "tungsten blue" ($[WO_2]_2WO_4$).

Molybdenum's and tungsten's most stable compounds are in the "+6" oxidation state, especially in aqueous solution. Molybdenum trioxide (MoO_3) is a white solid. Tungsten trioxide (WO_3) is a yellow solid. Otherwise, the "+6" state tends to be represented in molybdenum by *molybdates,* compounds in which Mo is present as the MoO_4^{2-} ion; and in tungsten by *tungstates,* compounds in which W is present as the WO_4^{2-} ion. Both ions are analogous to the chromate ion, CrO_4^{2-}.

Molybdates and tungstates both exhibit the ability to polymerize, with the extent of polymerization dependent on the *pH,* or acidity, of the solution. At high pH (low acidity), only MoO_4^{2-} and WO_4^{2-} exist. However, a decrease in the pH of the solutions (higher acidity) leads to the formation of polymolybdates and polytungstates that have formulas ranging from $Mo_3O_{11}^{4-}$ to $Mo_{24}O_{78}^{12-}$ and WO_3^- to $W_{12}O_{41}^{10-}$.

TUNGSTEN'S USE IN MUNITIONS

For at least half a millennium, lead and its alloys have been the materials of choice for musket balls, cannonballs, and bullets. Lead has a relatively low melting point and is, therefore, easily molded into shapes to fit various firearms.

Although the ill effects of lead absorption into the human system had become widely acknowledged by 1950, its use in bullets for hunting and military applications was largely considered to be outside the need for regulation. Firing ranges (both indoor and outdoor), however, pose serious health risks to humans. Indoor firing ranges are lead-inhalation hazards to users, while lead ammunition used at outdoor ranges can leach into water supplies. Exposure leads to especially severe problems for children, who may experience retarded growth, brain and nervous system damage, headaches, or hearing problems. Adults most commonly notice problems from continued (chronic) exposure. These risks led the U.S. Army and Marine Corps to begin using bullets manufactured from tungsten-based alloys in the late 1990s.

Just as lead is toxic to humans, it is also lethal to birds and other wildlife. Hunters often leave gut piles or even entire carcasses of animals they have shot in the wilderness. These remains are usually consumed by a variety of *carrion* eaters and scavengers—vultures, ravens, eagles, bears, coyotes, and other animals. Ingesting lead from shot and bullets in this manner significantly increases *mortality* (death rate). Concern about mortality among other birds and animals—especially species on the Endangered Species List, like the California Condor—has recently resulted in programs conducted by game and fish departments and national parks to use lead substitutes like tungsten-nylon bullets. In October 2008, California governor Arnold Schwarzenegger signed legislation banning lead ammunition.

Tungsten may not be a viable solution, however. While originally believed to be nontoxic and also unlikely to move into groundwater, recent studies offer cautionary insights. Research indicates that tungsten intake may inactivate some enzymes and shows evidence of carcinogenic effects. Some tungsten-mining town populations show higher-than-average leukemia rates, which may be linked to inhalation of tungsten dust. At the Camp Edwards army base in Massachusetts, abnormally high tungsten levels discovered in the groundwater have prompted the army to halt use of tungsten-based bullets and to remove tons of soil that may be contaminated. In 2006, Massachusetts governor Mitt Romney banned the use of tungsten-nylon bullets in that state.

A safe level for tungsten in the human system is unknown. There are no Environmental Protection Agency (EPA) guidelines for limits. However, the EPA has listed tungsten as "an emerging contaminant of concern."

DISCOVERING THE HAZARDS OF CHROMIUM IN DRINKING WATER

Sometimes it takes a nonscientist to see what researchers do not have the time, funding, or inclination to investigate. Such was the famous case of a tiny desert town whose residents turned up at local clinics with illnesses and disabilities in numbers that were far above average. Just outside Hinkley, California, Pacific Gas & Electric (PG&E) had, in the 1950s and 1960s, deemed it safe to release water from the cooling towers at its natural gas pipeline station into unlined holding ponds.

Normally, soil is an excellent filter against toxins, and over time, as the water seeps into *aquifers,* it can render some contaminated water safe for drinking. Such is not the case, however, for water that contains high amounts of hexavalent chromium, as was argued successfully by environmental activist Erin Brockovich in 1996. Universal Studios and director Steven Soderbergh turned the story into an Oscar-winning film (released in 2000) named after her.

Hexavalent chromium, also known as Cr (VI), is a component of certain rust-inhibiting paints, such as the paint that was used on the walls of the PG&E cooling towers. Unfortunately, some of that paint leached into the water, which subsequently seeped into the town's aquifers. While it was known that Cr (VI) is hazardous to inhale, little was understood about its effects when ingested. A National Toxicology Report released in 2007 by the National Institutes of Health indicates that new limits may need to be considered for drinking-water content. According to Dr. Michelle Hooth, who worked on the project, "Previous studies have shown that hexavalent chromium causes lung cancer in humans in certain occupational settings as a result of inhalation exposure. We now know that it can also cause cancer in animals when administered orally." The current upper limit prescribed by the Environmental Protection Agency for chromium in drinking water is 0.1 parts per billion (ppb). In August of 2009, the California Office of Environmental Health Hazard Assessment proposed a public health goal for state water supplies not to exceed 0.06 ppb.

TECHNOLOGY AND CURRENT USES OF THE CHROMIUM GROUP

Chromium has several familiar applications that include chrome plating and the production of stainless steel. A side effect of adding Cr or Mo to stainless steel is that the material becomes ferromagnetic. There are a number of important alloys that utilize molybdenum and tungsten. Tungsten's best-known application is its use as the filament in incandescent lightbulbs. These and other uses of these elements are summarized here.

Chromium is used extensively for *electroplating,* both to reduce wear and to enhance decorative effects by providing a bright surface. Examples include automobile parts and plumbing fixtures. Chromium is added to steel to make it resistant to corrosion. It is also used in ceramics for coloring. For many years, lead chromate was an important yellow pigment in paints. In recent years, however, because of health concerns associated with both lead and chromate, safer compounds have been substituted.

The major use of molybdenum is its addition to steel to increase hardness, toughness, and corrosion resistance. Particularly important is the use of molybdenum steels in tools, rifle barrels, and lightbulb filaments. Molybdenum steels are also used in food handling, hospital and laboratory equipment, automobile parts, and gas turbine parts. Molybdenum is also used in X-ray tubes, radio transmitting tubes, and electric furnaces. Compounds of molybdenum are used as paint pigments, catalysts in the petroleum refining industry, and flame-retardant materials. Molybdenum isotope data from ocean sediments are helpful to scientists attempting to decipher how much dissolved oxygen existed in the mid-*Proterozoic* oceans.

In addition to tungsten's application in the filaments of incandescent lightbulbs, tungsten carbides are important components of lathe tools. Tungsten alloys are used in the aerospace and defense industries, particularly in the manufacture of jet engine parts. Tungsten alloys are also used in surgical instruments and have been used to replace lead in ammunition. Compounds of tungsten are used in paints, dyes, and glass. A new polymer doped with molybdenum and tungsten may

Tungsten lightbulbs rely on resistance in the filament for light, but this also produces waste heat. *(Monkey Business Images/ Shutterstock)*

enhance the efficiency of *photovoltaic cells* by extending the range of wavelengths that can be utilized for solar power.

To conserve energy, society is replacing many of its incandescent lightbulbs with *fluorescent* bulbs. Otherwise, the many current uses for chromium, molybdenum, and tungsten are likely to continue and expand.

THE ASTROPHYSICS OF THE MANGANESE GROUP: Mn, Tc, Re

While members of the manganese group may be chemically similar, differences in astrophysical abundances make for ongoing research regarding their nucleosynthesis. Technetium and rhenium, both being heavier than iron, are synthesized in supernovae via the rapid capture by iron nuclei of a succession of neutrons, which is called the r-process. They are quickly blown out into space, where some of these heavy atoms are collected during the formation of new stars. Manganese, however, can build up slowly over thousands of years in the atmospheres of large-mass stars via neutron capture, with the requirement that iron 56 nuclei

THE BASICS OF MANGANESE

Symbol: Mn
Atomic number: 25
Atomic mass: 54.938
Electronic configuration: $[Ar]4s^23d^5$

T_{melt} = 2,275°F (1,246°C)
T_{boil} = 3,742°F (2,061°C)

Abundance in Earth's crust = 1,100 ppm

Manganese	
2	1246°
8	2061°
13	
2	

Mn 25

+2 +3 +4 +7
54.938049
0.000031%

Isotope	Z	N	Relative Abundance
$^{55}_{25}$Mn	25	30	100%

THE BASICS OF TECHNETIUM

Symbol: Tc
Atomic number: 43
Atomic mass: (All isotopes are radioactive with relatively short half-lives.)
Electronic configuration: $[Kr]5s^24d^5$

T_{melt} = 3,915°F (2,157°C)
T_{boil} = 7,709°F (4,265°C)

Abundance: not found in Earth's crust

Technetium	
2	2157°
8	4265°
18	
13	
2	

Tc 43

+4 +6 +7
[98]

Isotope	Z	N	Half-life
$^{95}_{43}$Tc	43	52	20.0 hours
$^{97}_{43}$Tc	43	54	2.60 million years
$^{98}_{43}$Tc	43	55	4.20 million years
$^{99m}_{43}$Tc*	43	56	6.0 hours
$^{99}_{43}$Tc	43	56	213 thousand years

Note: * = a *metastable* isotope

THE BASICS OF RHENIUM

Symbol: Re
Atomic number: 75
Atomic mass: 186.207
Electronic configuration:
  [Xe]$6s^2 4f^{14} 5d^5$

T_{melt} = 5,767°F (3,186°C)
T_{boil} = 10,105°F (5,596°C)

Abundance in Earth's crust = 0.003 ppm

Rhenium		
2		3186°
8	**Re**$_{75}$	5596°
18		
32		
13		
2	+4 +6 +7	
	186.207	
	1.69×10^{-10}%	

Isotope	Z	N	Relative Abundance
$^{185}_{75}$Re	75	110	37.40%
$^{187}_{75}$Re	75	112	62.60%

be available as seeds—remnants of prior supernova explosions. Because this synthesis proceeds relatively slowly due to a low density of neutrons, it is called the s-process.

The so-called manganese stars display spectra that indicate an unusually high ratio of manganese to iron in their stellar atmospheres. These stars tend to be much hotter and more massive than the Sun and rotate slowly. One theory is that the slow rotation rate may inhibit mixing of the manganese atoms once they are formed.

The discovery of technetium 99 in the spectra of *red giant* stars was a surprise because this isotope has a radioactive half-life of only 200,000 years. Unless recently fabricated, all Tc-99 should have decayed to ruthenium 99 in red giants, which are billions of years old. Studies of stardust grains have also provided evidence that Tc-99 must have been present when they formed. Astrophysicists have struggled with the question of how technetium is produced in stellar interiors. Could there be a slow neutron process that leads to its synthesis?

It turns out that a photon process may contribute to the technetium abundance, at least in asymptotic giant branch stars. The decay chain of zirconium 99 leads to Tc-99, as shown on page 59, with half-lives in parentheses.

$$^{99}\text{Zr}(35\text{ s}) \rightarrow {}^{99}\text{Nb}(2.4\text{ m}) \rightarrow {}^{99}\text{Mo}(67\text{ h}) \rightarrow {}^{99}\text{Tc}(2.1 \times 10^5\text{ y})$$

To initiate the fission event for zirconium, a photon of about 6 MeV is sufficient. In stars three to four times as massive as the Sun, carbon is produced by fusion of helium nuclei. Once carbon is made, a process called the *carbon-nitrogen-oxygen* (CNO) *cycle,* can take place in the following steps:

$$^{12}_{6}\text{C} + {}^{1}_{1}\text{H} \rightarrow {}^{13}_{7}\text{N} + \text{energy}$$

$$^{13}_{7}\text{N} \rightarrow {}^{13}_{6}\text{C} + e^+ + \nu$$

$$^{13}_{6}\text{C} + {}^{1}_{1}\text{H} \rightarrow {}^{14}_{7}\text{N} + \text{energy}$$

$$^{14}_{7}\text{N} + {}^{1}_{1}\text{H} \rightarrow {}^{15}_{8}\text{O} + \text{energy}$$

$$^{15}_{8}\text{O} \rightarrow {}^{15}_{7}\text{N} + e^+ + \nu$$

$$^{15}_{7}\text{N} + {}^{1}_{1}\text{H} \rightarrow {}^{12}_{6}\text{C} + {}^{4}_{2}\text{H} + \text{energy}$$

The energy produced is in the form of photons. In the third and fourth steps of this cycle, the photons carry energies of 7.55 MeV and 7.293 MeV, respectively—enough to initiate the zirconium radioactive decay chain, resulting in observable technetium 99.

DISCOVERY AND NAMING OF MANGANESE, TECHNETIUM, AND RHENIUM

The discovery of manganese is credited to the Swedish chemist Johan Gottlieb Gahn (1745–1818). Gahn's father died when Johan was just a young boy, forcing Johan to go to work in the local iron mines to support his family. It was there that he developed an interest in minerals and their analysis. Gahn was fortunate enough to have been able to study mineralogy under the Swedish chemist Torbern Bergman (1735–84). Manganese-containing minerals had been used for 200 years to color glass with a violet color. However, no one had analyzed the minerals to determine their composition. Bergman tried analyzing a mineral that was called black magnesia because it was mistakenly thought to contain magnesium. Bergman succeeded in showing that it did not contain magnesium but probably contained a new element. He was unsuccessful, however, in his attempts to isolate it.

Bergman turned the problem over to his friend Carl Scheele. Scheele concluded that a new element was present, which he called manganese, but he, too, was unable to isolate it. Gahn then continued the investigation. In 1774, he succeeded in reducing the manganese ore to pure metallic manganese. There are two possible origins of the name *manganese*. One possibility is that the name was derived from the Latin word for magnet *(manges)* because one ore of manganese is slightly magnetic. The other possibility is that the name comes from *black magnesia*.

Probably no one in the 18th century would have suspected it would take so long, but the discoveries of the remaining Group VIIA metals did not occur until the 20th century. In fact, rhenium, which was isolated in 1925, was the last naturally occurring element to be discovered. Technetium was not found in nature, but had to be produced artificially.

At the beginning of the 1920s, chemists realized that elements 43 and 75 were the only two transition elements yet to be discovered. It was known that the elements chromium to copper (numbers 24–29), ruthenium to silver (numbers 44–47), and osmium to gold (numbers 72–74) occur in association with other platinum metals. In addition, it was known that elements yttrium to molybdenum (numbers 39–42) and elements hafnium to tungsten (numbers 72–74) are found in deposits of columbite. It seemed reasonable, therefore, to look for the missing element 43 in platinum group ores and for element 75 in deposits that contained columbite. It was also reasonable to assume that elements 43 and 75 would have chemical properties that were similar to those of manganese.

In 1925, the German chemists Walter Noddack (1893–1960), Ida Tacke (1896–1978), and Otto Berg (1906–91) carefully studied the X-ray emission lines of all the transition elements and reported the discovery of element 75. The X-ray lines attributed to the element did not match the lines of any of the other known elements. They chose to name the new element *rhenium* in honor of the Rhine River, which is a major river in Germany.

In addition, using X-ray analysis, Noddack, Tacke, and Berg reported the discovery of element 43, which they named *masurium*. Subsequent investigations, however, failed to confirm element 43's discovery. After their joint discovery of rhenium, Noddack and Tacke married and continued to conduct scientific research together.

Element 43 was the first element to be discovered that does not occur naturally on Earth. Element 43 actually has a history that predates the work of Mendeleev or any use of the periodic table to predict the element's existence. Beginning in the mid-1850s, various investigators had claimed the element's existence and put forth names for it. Some of the names suggested for an element that would lie between molybdenum and ruthenium were *davyum* (for Humphrey Davy), *nipponium* (Nippon being the Japanese name for Japan), *moselium* (for Henry Moseley), *masurium* (for the Masurian marshes in eastern Prussia), and *trinacrium* (Trinacria being a name for Sicily).

The attempt to find element 43 was resumed in the 1930s by the Italian physicist Emilio Segrè (1905–89). In the 1930s, American physicist Ernest Orlando Lawrence (1901–58) of the University of California at Berkeley invented the *cyclotron,* a particle accelerator in which elementary particles can be accelerated to relativistic speeds and made to smash into other target particles. The result can be the creation of new elementary particles, of different isotopes of the target atoms, or of new elements altogether. (Lawrence received the Nobel prize in physics in 1939 for the invention of the cyclotron.) In 1936, Lawrence bombarded a molybdenum target with *deuterons* (heavy nuclei of hydrogen) for several months. He sent the resulting substance to Segrè and chemist Carlo Perrier (1886–1948) in Italy for analysis. The two scientists showed that the substance that had been formed was highly radioactive, very similar chemically to rhenium, and less similar chemically to manganese.

Despite the fact that Segrè and Perrier had only 10^{-10} gram of the new substance with which to work, they concluded that a new element—number 43—had in fact been synthesized. Because all of its isotopes are radioactive, with half-lives significantly shorter than the age of Earth, element 43 does not occur naturally on Earth. Later, in 1940, larger amounts of element 43 were found among the products of the nuclear fission of uranium.

Even after confirmation of element 43's discovery, at first it was not clear what status to give an artificial element, or whether or not it even deserved a name. In 1947, however, it was decided that an artificial element is an element, nonetheless, and that the discoverers have the right to give it a name. Segrè and Perrier chose the name *technetium* from the Greek word for *technical* (because it was produced by technical, or artificial, means) and gave it the symbol Tc.

THE CHEMISTRY OF THE MANGANESE GROUP

Manganese is element 25, with a density of 7.4 g/cm³, and ranks 12th in abundance on Earth. Manganese is mined in several countries around the world including Australia, Brazil, Gabon, India, Russia, and South Africa. Because ores in the United States have very low manganese contents, it is more economical to import manganese than to recover it domestically. Usually, manganese is found as the black or gray mineral manganite (MnO[OH)]). A potential source of manganese is the ocean floor, where black nodules of pyrolusite (manganese dioxide, MnO_2) have been found. Presently, however, these nodules are usually too deep to make recovery economically feasible. Lesser sources include rhodochrosite (manganese carbonate, $MnCO_3$), rhodonite (a silicate that also contains iron, magnesium, and calcium), and bixbyite (an oxide that also contains iron). Manganese is an essential element for animals: It increases the strength of bones, aids in absorbing vitamin B1, and is an important *cofactor* for enzymes.

As mentioned previously, technetium—element 43—is not found naturally on Earth. All of its isotopes are radioactive, with half-lives significantly shorter than the age of Earth, so that any *primordial* technetium that may have been present when Earth first formed has long since decayed away. However, appreciable quantities of technetium are made in nuclear reactors. Technetium has a density of 11.5 g/cm³. The chemistry of technetium has not been studied extensively, but it closely resembles rhenium in the kinds of compounds that it forms. Of greater importance is the use of the metastable isotope Tc-99, which is a gamma-ray emitter used in nuclear medicine.

Rhenium is element 75, with a very high density of 21.0 g/cm³, and ranks 77th in abundance on Earth. Worldwide production of rhenium has always been quite low. Beginning in 1935, commercial quantities of rhenium began to be recovered from *potash* mines in Germany. That source eventually became exhausted and no longer produces rhe-

A potential source of manganese is the ocean floor, where black nodules of pyrolusite (manganese dioxide) have been found. *(Charles D. Winters/Photo Researchers, Inc.)*

nium. In 1957, rhenium began to be recovered from molybdenum sulfide, which is a by-product of copper mining in Arizona, especially at the Phelps Dodge copper mine in the town of Bagdad. According to the United States Geological Survey (USGS), annual world mine production of rhenium is about 43 tons. The major producers are Chile, Kazakhstan, the United States, and Peru.

Since manganese, technetium, and rhenium are in Group VIIA, the "+7" oxidation state would be expected to be important, and, in fact, it is. Compounds of manganese and rhenium are known in which these elements exist in positive oxidation states ranging from "+2" to "+7". In the "+7" state, for example, the elements form the compounds Mn_2O_7, which is red in color, and Re_2O_7, which is yellow in color. In addition, rhenium has been observed to exist in a "−1" oxidation state, which is exceptionally unusual for a metal. Technetium compounds have been studied mainly in the "+7" state.

Manganese metal is readily oxidized in the presence of oxygen gas, water, and hydrochloric acid. For example, in hydrochloric acid, the following reaction takes place:

$$Mn \text{ (s)} + 2 \text{ HCl (aq)} \rightarrow MnCl_2 \text{ (aq)} + H_2 \text{ (g)}.$$

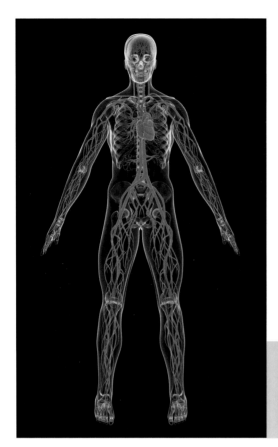

Technetium 99 can be used to map circulatory system disorders. *(Sebastian Kaulitzki/ Shutterstock)*

In compounds, manganese can be in any of the oxidation states from "+2" to "+7," inclusive. Examples of manganese in these states are illustrated in the following series:

Oxidation state	0	"+2"	"+3"	"+4"	"+5"	"+6"	"+7"
Chemical species	Mn	Mn^{2+}	Mn^{3+}	MnO_2	MnO_3^-	MnO_4^{2-}	MnO_4^-.

The manganic ion (Mn^{3+}) tends to be unstable in aqueous solution and disproportionates into Mn(II) and Mn(IV), as shown in the following reaction:

$$2\ Mn^{3+}\ (aq) + 2\ H_2O\ (l) \rightarrow Mn^{2+}\ (aq) + MnO_2\ (s) + 4\ H^+\ (aq).$$

Disproportionation means that an element initially in only one oxidation state changes to two products—one product exhibiting a lower oxidation state and the other exhibiting a higher state.

The oxides formed in the "+2" and "+3" states—manganous oxide (MnO) and manganic oxide (Mn_2O_3), respectively—are bases due to their solubility in acid solutions, as shown in the following example:

$$MnO \ (s) + 2 \ HCl \ (aq) \rightarrow MnCl_2 \ (aq) + H_2O \ (l).$$

In the "+4" state, manganese forms black manganese dioxide (MnO_2), which is the only compound of manganese (IV). Manganese dioxide is amphoteric because it is soluble in both acid and basic solutions. In higher oxidation states, oxides of manganese are acidic because they dissolve in basic solutions.

The manganate ion (MnO_4^{2-}) disproportionates in acidic solution to the "+4" and "+7" states, as shown in the following reaction:

$$3 \ MnO_4^{2-} \ (aq) + 4 \ H^+ \ (aq) \rightarrow MnO_2 \ (s) + 2 \ MnO_4^- \ (aq) + 2 \ H_2O \ (l),$$

where MnO_4^- is the purple-colored permanganate ion in which manganese is in the "+7" oxidation state.

In acidic solution, permanganate also oxidizes the manganous ion (the "+2" state) to form manganese dioxide (the "+4" state), as shown in the following equation:

$$2 \ MnO_4^- \ (aq) + 3 \ Mn^{2+} \ (aq) + 2 \ H_2O \ (l) \rightarrow 5 \ MnO_2 \ (s) + 4 \ H^+ \ (aq).$$

In this reaction, manganese begins in two different oxidation states and forms a product that exhibits an oxidation state intermediate between the two initial states.

As mentioned previously, very little study has been done of the chemistry of technetium. At least three oxides are known: TcO_2, which is black; TcO_3, which is purple; and Tc_2O_7, which is yellow. In these compounds, technetium is in the "+4," "+6," and "+7" oxidation states, respectively. Technetium also exists in the "+7" state in the pertechnetate ion (TcO_4^-), which is readily reduced by copper metal to metallic technetium, as shown in the following reaction taking place in acidic solution:

$$2\ TcO_4^- \text{ (aq)} + 7\ Cu \text{ (s)} + 16\ H^+ \text{ (aq)} \rightarrow$$
$$2\ Tc \text{ (s)} + 7\ Cu^{2+} \text{ (aq)} + 8\ H_2O \text{ (l)}.$$

Technetium also forms halides. $TeCl_4$ is deep red, $TeCl_6$ is green, and TeF_6 is yellow.

Rhenium's chemistry is very similar to the chemistry of manganese. Rhenium forms colored oxides in several oxidation states, as shown in the following table:

OXIDATION STATE	FORMULA OF OXIDE	COLOR OF OXIDE
"+3"	Re_2O_3	Black
"+4"	ReO_2	Brown
"+5"	Re_2O_5	Blue
"+6"	ReO_3	Red
"+7"	Re_2O_7	Yellow

Rhenium also forms a number of colored halides, as shown in the following table:

OXIDATION STATE	FORMULA OF HALIDE	COLOR OF HALIDE
"+2"	ReI_2	Black
"+3"	Re_3Cl_9	Red
	Re_3Br_9	Red-Brown
	Re_3I_9	Black
"+4"	ReF_4	Blue
	$ReCl_4$	Black
	$ReBr_4$	Red
	ReI_4	Black
"+5"	ReF_5	Green-Yellow
	$ReCl_5$	Red-Brown
	$ReBr_5$	Green-Blue
"+6"	ReF_6	Yellow
	$ReCl_6$	Green-Brown
"+7"	ReF_7	Yellow

In analogy to the permanganate (MnO_4^-) and pertechnetate (TcO_4^-) ions, rhenium forms the perrhenate ion (ReO_4^-). In acidic solution, all three ions are powerful oxidizing agents. In practice, however, because of cost, MnO_4^- is the only ion of the three that would commonly be used in the laboratory. MnO_4^- usually is purchased in the form $KMnO_4$, a deep-purple crystalline solid.

TECHNOLOGY AND CURRENT USES OF THE MANGANESE GROUP

Although the elements in the manganese group are less familiar to most people than the elements in the chromium group, they do have several important applications.

Manganese is an essential component of most forms of steel: Examples of applications are I-beams in building construction, naval armor plate, bulldozer blades, and dredger buckets. Manganese may also be used to strengthen brass and to color bricks. In addition, a form of bronze, called manganese bronze, contains about 60 percent copper, about 35 percent zinc, and 1 to 4 percent manganese.

Manganese dioxide is one of the most common manganese compounds and is used in dry-cell batteries and to decolorize window and bottle glass. Potassium permanganate ($KMnO_4$) is used to kill bacteria and algae in water and wastewater treatment. Potassium permanganate is also an important oxidizing agent in organic syntheses. Manganese sulfate ($MnSO_4$) is added to animal feeds and plant fertilizers as a source of manganese.

Technetium's radioactivity both limits its use and provides its most valuable uses. The main isotope of technetium, Tc-99, has important applications in radiology, particularly in diagnostic tests. The use of technetium 99 in nuclear medicine saves lives every day.

Rhenium has numerous uses as a catalyst. Of particular current interest is the possibility that it can initiate a reaction separating hydrogen from water, a function that could prove valuable in hydrogen fuel development. Furthermore, the addition of rhenium to metals like molybdenum and tungsten increases the ductility and *tensile strength* of those metals. Rhenium alloys are also used in semiconductors, thermocouples, nuclear reactors, *gyroscopes,* and electrical contacts.

4

The Iron, Cobalt, and Nickel Groups

The iron group consists of iron (Fe, element 26; *ferrum* in Latin), ruthenium (Ru, element 44), and osmium (Os, element 76). The cobalt group consists of cobalt (Co, element 27), rhodium (Rh, element 45), and iridium (Ir, element 77). The nickel group consists of nickel (Ni, element 28), palladium (Pd, element 46), and platinum (Pt, element 78).

In the preceding chapters, vertical triads of elements have been grouped together for discussion, and this is how the IUPAC officially classifies a "group." A shift occurs, however, after the manganese group elements. The next three columns of elements are all clustered together as Group VIIIA elements (in the original Roman-numeral labeling system). The reason for this shift was that, with the Group VIIIA elements, horizontal triads are more important than vertical ones. In other words, instead of discussing iron, ruthenium, and osmium as a triad, it makes more sense chemically to group iron, cobalt, and nickel as a triad. The properties of those three elements resemble each other much

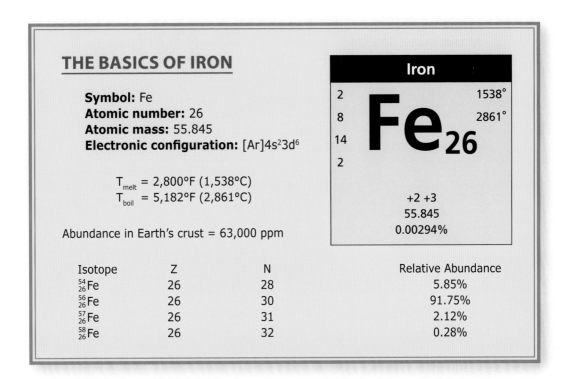

THE BASICS OF IRON

Symbol: Fe
Atomic number: 26
Atomic mass: 55.845
Electronic configuration: $[Ar]4s^23d^6$

T_{melt} = 2,800°F (1,538°C)
T_{boil} = 5,182°F (2,861°C)

Abundance in Earth's crust = 63,000 ppm

Iron		
2		1538°
8	**Fe**$_{26}$	2861°
14		
2		
	+2 +3	
	55.845	
	0.00294%	

Isotope	Z	N	Relative Abundance
$^{54}_{26}$Fe	26	28	5.85%
$^{56}_{26}$Fe	26	30	91.75%
$^{57}_{26}$Fe	26	31	2.12%
$^{58}_{26}$Fe	26	32	0.28%

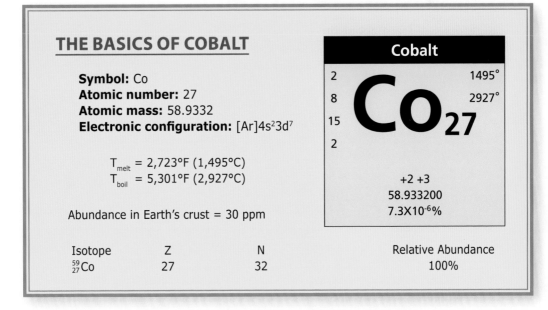

THE BASICS OF COBALT

Symbol: Co
Atomic number: 27
Atomic mass: 58.9332
Electronic configuration: $[Ar]4s^23d^7$

T_{melt} = 2,723°F (1,495°C)
T_{boil} = 5,301°F (2,927°C)

Abundance in Earth's crust = 30 ppm

Cobalt		
2		1495°
8	**Co**$_{27}$	2927°
15		
2		
	+2 +3	
	58.933200	
	7.3×10^{-6}%	

Isotope	Z	N	Relative Abundance
$^{59}_{27}$Co	27	32	100%

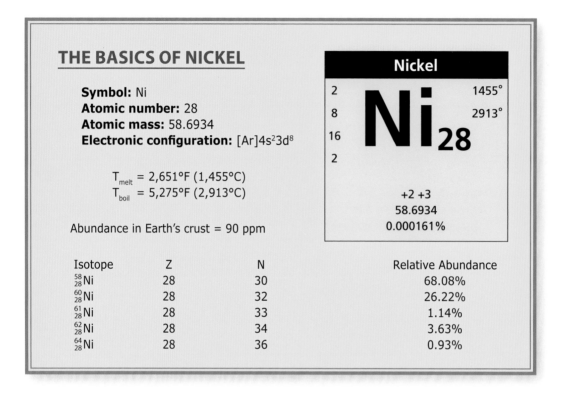

THE BASICS OF NICKEL

Symbol: Ni
Atomic number: 28
Atomic mass: 58.6934
Electronic configuration: $[Ar]4s^2 3d^8$

T_{melt} = 2,651°F (1,455°C)
T_{boil} = 5,275°F (2,913°C)

Abundance in Earth's crust = 90 ppm

Nickel

2
8
16
2

Ni 28

1455°
2913°

+2 +3
58.6934
0.000161%

Isotope	Z	N	Relative Abundance
$^{58}_{28}$Ni	28	30	68.08%
$^{60}_{28}$Ni	28	32	26.22%
$^{61}_{28}$Ni	28	33	1.14%
$^{62}_{28}$Ni	28	34	3.63%
$^{64}_{28}$Ni	28	36	0.93%

more closely than they do the elements lying below them. Similarly, ruthenium, rhodium, and palladium form a triad, as do osmium, iridium, and platinum. Together, these latter two triads are classified as the platinum metals because of their similar properties. In this chapter, the reader will learn about the chemical and physical properties of iron, cobalt, nickel, and the platinum metals.

Iron, cobalt, and nickel are similar in many ways. First of all, they all tend to react readily with common acids. (In contrast, the platinum metals exhibit *noble* behavior, which means that they tend *not* to react with even strong reagents like hydrochloric or nitric acid.) Second, iron, cobalt, and nickel all form compounds in the "+2" oxidation state: Halides, oxides, nitrates, sulfides, and sulfates of all three "+2" ions are quite common. Third, there is a decreasing stability of the "+3" oxidation state. Iron readily forms a "+3" ion and exists in the "+3" state in several compounds. Cobalt, however, does not form a free "+3" ion, but only exists in the "+3" state in certain *complex ions.* For all practical purposes, nickel does not exhibit a "+3" oxidation state at all. Fourth, iron, cobalt, and nickel are all very hard metals with high melting and boiling points. Fifth, all three metals exhibit magnetic properties. In fact, the term used by scientists to describe permanent magnetic behavior in materials is *ferromagnetism,* where the prefix *ferro* is a reference to iron.

Iron, and its most important alloy, steel, is found everywhere. Even young children are familiar with iron and steel, from the cartoons depicting iron in Popeye's spinach to the "man of steel," Superman. Iron is an essential component of the hemoglobin that carries oxygen through the blood stream. Cars, trucks, locomotives, bridges, skyscrapers, oceangoing vessels, nails, bolts, tanks, armor, and weapons are all made at least in part of steel. The Iron Age is the age of modern civilization. The Industrial Revolution was built with steel. Iron even plays a crucial role in cosmology. Iron nuclei are the most stable nuclei of all the elements: Only the most massive and hottest of stars are able to synthesize elements heavier than iron.

THE ASTROPHYSICS OF IRON, COBALT, AND NICKEL

The fusion of hydrogen, helium, and other light elements in stellar cores generates energy in the form of radiation. This radiation, emitted outward from the center, counteracts the force of gravity and keeps the star from collapsing under its own weight. In stars heavier than about 10 solar masses, the fusion of lighter elements eventually produces iron at the core—which also harbors elementary particles such as electrons, positrons, protons, neutrons, and photons. Iron will not fuse without energy input, so *electron pressure* must support the star, but hydrogen-forming collisions of electrons with protons deplete the electron population and, therefore, the electron pressure to the point that it is no match for gravity.

When gravity becomes the dominant force, the star contracts with increasing velocity. Material in the outer envelope contracts faster than material at smaller distances from the center, and outer material can reach *supersonic* speeds. The core temperature rises fast during the collapse, producing gamma rays that are energetic enough to break iron nuclei into free-flying helium nuclei, protons, neutrons, and electrons in a process called *photodisintegration.* The electrons and protons can then combine to form neutrons and *neutrinos*. The neutrinos are so light and noninteracting that they immediately fly away from the core while the number of neutrons continues to rise and compactify under the force of gravity.

The entire process is not yet well understood, but there comes a point at which *neutron degeneracy pressure* will not allow further compactification of the star. At that instant, the *infalling* material rebounds off the newly stable neutron core and travels back outward. When this material collides with the supersonic outer envelope matter that is still falling inward, a shock wave of incredible proportion results. This is the supernova event, during which almost all elements heavier than iron are synthesized.

Current supernova theory indicates that iron should be ejected from the star during the process, but as of September 2009, such iron *ejecta* have only been observed in one supernova remnant. In this object, known as G11.2-0.3, spectra show jets of iron matter exiting the neutron core in opposing (back-to-back) directions. This is surprising, since many astrophysicists expected supernova explosions to have spherical symmetry. Other curiosities, such as how the shock wave could sustain itself, will keep supernova research active for the foreseeable future.

Other ongoing astrophysical research on iron is focused on the iron-to-oxygen abundance ratio in *blue compact galaxies*. In this type of galaxy, iron appears to be underproduced with respect to oxygen (as compared with the solar ratio), displaying a relative abundance similar to that in *galactic halo* stars. More observational data will probably need to be examined before this phenomenon is understood, though it may be related to early supernova activity in the vicinity of the galactic halo.

Supernova explosions that occur in the manner described here are called Type II supernovae (SNe II). Another type of supernova event—a Type I supernova—can occur in two-star systems where one member is an extremely dense, compact, cool star called a *white dwarf*. As the two stars revolve around their common center of mass, gases from the outer region of the companion star are gravitationally attracted to the white dwarf. Eventually, the white dwarf acquires so much matter from the companion star that the infalling material begins to fuse. As in SNe II, there is a maximum density beyond which the star cannot compact any further. When that density is reached, the additional energy from the continually infalling matter results in a violent thermonuclear explosion in which the entire star is blown apart. At the moment of the explosion, there is enough energy within the core to allow iron 52 to fuse with helium nuclei (alpha particles), creating large quantities of nickel 56, as seen in the following reaction:

$$^{52}_{26}\text{Fe} + ^{4}_{2}\alpha \rightarrow ^{56}_{28}\text{Ni}.$$

The nickel 56 atoms effectively light the supernova for days to months, depending on how much nickel is formed. Each atom of nickel 56 decays, with a half-life of 6 days, to cobalt 56 and a gamma ray. Cobalt 56 atoms decay with a half-life of 77 days, producing iron 56 and more gamma rays. The gamma rays interacting with the rest of the exploding matter cause it to glow with what was once believed to be a standard *luminosity*. That is, the light given off by Type I supernovae (SNeI), which occur at the rate of about one per second, has been considered so uniform that they have been used as a *standard candle* for research into the expansion of the universe. This seemed plausible because it was believed that the limiting mass, beyond which a white dwarf would explode, was 1.4 times the mass of the Sun (the *Chandrasekhar mass*). If all Type I supernovae begin with the same mass and the same elements at the core, they should all have the same brightness. In 2003, however, astrophysicists observed a superbright SNeI, that expanded at a surprisingly slow rate, which seemed to indicate that it had more gravitational force holding it back. If that is the explanation, it would have to be about 50 percent more massive than the Chandrasekhar limit. That would be possible if the star were rotating before it exploded, and probably many white dwarfs do rotate.

Researchers are currently revisiting their theories on the formation and evolution of these important markers of the universe. In 2008, French scientists reported the first successful human synthesis of a nickel 56 nucleus, which requires extreme compression. Further studies of this process will help with an understanding of supernovae and neutron stars.

DISCOVERY AND NAMING OF IRON, COBALT, AND NICKEL

The discoverer of iron is unknown. Ancient people recognized meteorites as a source of iron as early as 3500 B.C.E. and made a limited number of weapons and tools from it. Presumably, sometime in the interval 3000–2500 B.C.E., charcoal in a campfire was observed to reduce iron oxide in the soil to bits of elemental iron. The iron thus produced was recognized for its strength, and inhabitants of the ancient Mediterranean, Near Eastern, and Far Eastern cultures began to fashion implements from it. Spearheads and knives were made of iron, as were chariots and city gates. Fishermen began making fishhooks from iron instead of from bones. Carpentry tools and cooking utensils were made from iron. The blacksmith trade was created.

By 1000 B.C.E., ironsmiths in the Middle East were making steel. Magnetite, or lodestone (Fe_3O_4), was discovered in the Mediterranean world during the sixth century B.C.E. and its magnetic properties recognized. The Latin name for iron, *ferrum*, dates from this era and is the source of the symbol for iron, Fe. (The modern name *iron* is an Anglo-Saxon term.)

Ancient iron artifacts, such as these spearheads, have been found in nearly every country. *(Protohistoric/ The Bridgeman Art Library/Getty Images)*

THE IRON AGE

The Iron Age is loosely dated from the time when civilizations made the transition from using bronze for most tools and weapons (the Bronze Age) to using mostly iron (the exception being in Africa, where bronze was not worked). The onset of the Iron Age was different for different regions, but the reasons for the switch from bronze to iron were probably the same everywhere. First, iron ore is more plentiful in Earth's crust than copper and tin—the components of bronze. Second, pure iron is stronger than bronze, and steel is even stronger.

Ornaments and spear tips made of iron originating from as long ago as 4000 B.C.E. have been found in areas of ancient Egypt and Mesopotamia. Analysis shows a small nickel content in these iron pieces, which suggests that they were formed from pieces of meteorite, rather than ore that was mined and worked. The oldest iron implement created by melting and hammering ore was found in Egypt and dates to 1350 B.C.E., but it seems to have originated in Anatolia (modern-day Turkey), where the Hittite civilization was quite advanced in iron working, and may have kept their methods to themselves until the fall of their empire in 1200 B.C.E. Around this time, regions that had entered the Iron Age included India, modern-day Iran, and West Africa. People in Niger were among the first to use *smelters* to extract the pure metal from iron-containing ore, perhaps as early as 1500 B.C.E. Farming in the area became widespread with the advent of iron tools, and the region became quite wealthy and powerful—a stark contrast to current conditions in Niger, which is mostly desert. By 900 B.C.E., Europe had entered the Iron Age, and artisans in China were casting iron into molds to make all manner of weapons and ornaments. Iron-working techniques spread throughout Europe by about 600 B.C.E., leading to great advances in nearly every area of life.

Iron smelters, however, required huge amounts of fuel in the form of charcoal, and centuries of feeding smelters eventually led to large-scale deforestation. Coal was not a viable fuel because

(continues)

(continued)

of its many impurities, but around 1700 C.E., after much trial and error, the Darby family in England found a way to remove impurities from coal to make a usable smelting fuel called *coke*. The result is a highly carbonized iron known as *pig iron,* which is rather brittle, but can be worked into *wrought iron*. A small amount of carbon in iron (up to 2 percent) actually strengthens it. Steel is iron alloyed with carbon, produced by burning off the excess carbon in pig iron. By the mid-19th century, steel was in mass production, heralding the Industrial Revolution. The Iron Age continues to this day, as so many everyday items contain steel.

Rock containing iron ore has been mined for millennia.
(Michael Steden/Shutterstock)

Cobalt and nickel were both discovered in Sweden. Cobalt ore had been used for centuries to give glass a blue color. It was not until the 1700s, however, that metallic cobalt was isolated and identified. Cobalt's discoverer was the Swedish chemist and mineralogist Georg

Brandt (1694–1768). Brandt's father was a mine owner. Together, father and son conducted elementary chemistry experiments. Brandt became intrigued by the blue color of some glasses. At the time, *smelters* of the ore thought the blue color was due to the presence of copper. However, they were unable to isolate any copper from the ore and became convinced that evil underground goblins (in German, *kobolds*) were interfering in their experiments. (Kobolds had such a bad reputation for sabotaging the work of miners that inhabitants of mining towns would pray in the churches for protection from them.) In 1735, when Brandt succeeded in isolating a new element that was not copper, he named it *cobalt* for the goblins. Cobalt was the first metal to have been discovered since ancient times.

Nickel was discovered in 1751 by the Swedish chemist Axel Fredrik Cronstedt (1722–65). Cronstedt studied mining and chemistry under Brandt. Nickel's history is similar to the history of cobalt. Like cobalt, nickel ores had also been known since ancient times, as was nickel's ability to color glass. Like cobalt, nickel was also difficult to isolate and at first was called *kupfernickel,* or false copper. When nickel was finally isolated, the word *nickel* (the German word for *satan*) was chosen for reasons similar to the choice of the word *cobalt.*

Cronstedt's claim of having discovered nickel was not universally accepted. In France, Balthasar-Georges Sage (1740–1824) and Antoine Grimoald Monnet (1734–1817) challenged Cronstedt's claim. They thought that Cronstedt's substance was not a pure element, but instead was a mixture of several metals. Torbern Bergman resolved the controversy by showing that other elements like cobalt and iron were minor impurities in Cronstedt's substance. After removing all of the impurities, Bergman demonstrated that he could obtain nickel of very high purity. Most people, although not everyone, became convinced of Cronstedt's claim and credited Cronstedt with the discovery of nickel.

THE CHEMISTRY OF IRON

Iron is element 26, with a density of 7.86 g/cm^3. It is the 4th most abundant element on Earth and the second most abundant metal (after aluminum). Iron's many uses, its significance in human history, and its important role in animal biology make iron one of the most familiar of all elements. There are several important iron ores spread widely throughout the world. Probably the most important ore is hematite (ferric oxide, Fe_2O_3), ranging in color from a familiar rusty red to steel-gray, or a metallic black color. A second major iron ore is magnetite (Fe_3O_4), the magnetic black iron oxide also known as *lodestone.* Goethite and limonite—which are forms of FeO(OH)—are both yellow, brown, or blackish iron oxides. Iron pyrite (FeS_2), or *fool's gold,* is a sulfide of iron with a pale yellow color. Meteorites are usually rich in metallic iron, hence the high density of meteorites. Because of iron's ubiquity in Earth's crust, iron is also a constituent of ores with mixed metals.

Iron is an essential element and is the major component of *hemoglobin,* the compound in blood that transports oxygen through the body. Iron is also an important nutrient essential for plant growth.

The chemistry of iron is dominated by the "+2" and "+3" oxidation states. These oxidation states are illustrated in the reactions that take place in a *blast furnace* when iron ore is reduced to iron metal in the production of steel. Different regions of the blast furnace are at different temperatures. In a region of relatively cooler temperature (about 400°F [200°C]), ferric oxide reacts with carbon monoxide (CO) according to the following chemical reaction:

$$3\ Fe_2O_3\ (s) + CO\ (g) \rightarrow 2\ Fe_3O_4\ (s) + CO_2\ (g).$$

In Fe_3O_4, iron is in both the "+2" and "+3" oxidation states, as can be understood by rewriting the formula of Fe_3O_4 as $FeO \cdot Fe_2O_3$. (In FeO, iron is in the "+2" state, while in Fe_2O_3 it is in the "+3" state.) At higher temperatures (about 660°F [350°C]), Fe_3O_4 is reduced to ferrous oxide (FeO) as shown in the following reaction:

$$Fe_3O_4\ (s) + CO\ (g) \rightarrow 3\ FeO\ (s) + CO_2\ (g).$$

At still higher temperatures, ferrous oxide is finally reduced to iron metal, as shown in the following reaction:

$$FeO\ (s) + CO\ (g) \rightarrow Fe\ (s) + CO_2\ (g).$$

Once metallic iron has been produced, it can be alloyed to make steel.

The most familiar oxide of iron is ferric oxide, which is the composition of rust. In a moist environment, metallic iron tends to corrode to ferrous oxide, as shown in the following reaction using atmospheric oxygen as the oxidizing agent:

$$2 \text{ Fe (s)} + \text{O}_2 \text{ (g)} \rightarrow 2 \text{ FeO (s)}.$$

Note, however, that the oxidizing agent could be water, weak acids found in the environment, or a combination of these and/or oxygen gas. Ferrous oxide tends to oxidize further to ferric oxide, as shown by the following reactions:

$$4 \text{ FeO (s)} + \text{O}_2 \text{ (g)} + 6 \text{ H}_2\text{O (l)} \rightarrow 4 \text{ Fe(OH)}_3 \text{ (s)};$$

$$2 \text{ Fe(OH)}_3 \text{ (s)} \rightarrow \text{Fe}_2\text{O}_3 \text{ (s)} + 3 \text{ H}_2\text{O (l)}.$$

Both Fe(OH)_3 and Fe_2O_3 have the familiar reddish-brown color of rust. What compound is present depends largely on the amount of moisture present.

The ferrous ion readily undergoes disproportionation reactions in which one ferrous ion is reduced to a neutral iron atom and two ferrous ions are oxidized to a ferric ion. This process is shown in the following reaction:

$$3 \text{ Fe}^{2+} \text{ (aq)} \rightarrow \text{Fe (s)} + 2 \text{ Fe}^{3+} \text{ (aq)}.$$

Because the ferrous ion is easily oxidized by atmospheric oxygen, aqueous solutions containing the ferrous ion tend to be unstable. If solutions of ferrous chloride (FeCl_2) or ferrous sulfate (FeSO_4), for example, are needed for laboratory work, they should be prepared fresh and used promptly.

Both Fe^{2+} and Fe^{3+} form complex ions. The most frequently encountered complex ions in high school or beginning college chemistry labs are the cyanides of these ions. The complex ions Fe(CN)_6^{4-} and Fe(CN)_6^{3-} often are found as the potassium salts $\text{K}_4\text{Fe(CN)}_6$ (potassium ferrocyanide) and $\text{K}_3\text{Fe(CN)}_6$ (potassium ferricyanide). Ferric ions also combine with the thiocyanate ion (SCN^-) to form a blood-red complex ion with the formula FeSCN^{2+}.

The following table lists several of the common ferrous and ferric compounds along with their colors:

	Cl^-	NO_3^-	O^{2-}	CO_3^{2-}	SO_4^{2-}	S^{2-}
Fe^{2+}	$FeCl_2$ (gray)	$Fe(NO_3)_2$ (gray)	FeO (white)	$FeCO_3$ (gray)	$FeSO_4$ (blue-gray)	FeS (black)
Fe^{3+}	$FeCl_3$ (yellow)	$Fe(NO_3)_3$ (gray)	Fe_2O_3 (red-brown)	$Fe_2(CO_3)_3$ (red)	$Fe_2(SO_4)_3$ (yellow)	Fe_2S_3 (green)

THE CHEMISTRY OF COBALT

Cobalt is element 27, with a density of 8.9 g/cm³, and ranks 32nd in order of abundance on Earth. The most important mineral that contains cobalt is cobaltite, which has a composition of CoAsS. There are reserves of about 1 million tons of cobalt in the United States, mostly in the state of Minnesota. Cobaltite is fairly rare, with deposits occurring mainly in Norway, Sweden, and Ontario, Canada. Cobalt is also found in meteorites that contain iron and nickel, in manganese nodules on the bottom of the ocean, and in association with other transition metals such as nickel.

Cobalt is a moderately active metal and dissolves in hydrochloric acid to form the "+2" ion, as shown in the following equation:

$$Co \; (s) + 2 \; HCl \; (aq) \rightarrow Co^{2+} \; (aq) + 2 \; Cl^- \; (aq) + H_2 \; (g).$$

The Co^{2+} ion is pink in color and is called the *cobaltous* ion; it is the only cobalt ion that occurs in simple compounds. Examples of Co^{2+} compounds include the following: cobaltous chloride ($CoCl_2$), cobaltous nitrate ($Co[NO_3]_2$), cobaltous sulfate ($CoSO_4$), cobaltous hydroxide ($Co[OH]_2$), cobaltous oxide (CoO), cobaltous carbonate ($CoCO_3$), cobaltous phosphate ($Co_3[PO_4]_3$), and cobaltous sulfide (CoS). The first three are soluble in dilute aqueous solution, whereas the last five are all insoluble. All of these compounds exhibit bright colors, usually red, blue, or black. Paint pigments include *cobalt blue,* which is cobalt aluminate ($Co[AlO_2]_2$), and cobalt green, which is cobalt zincate ($CoZnO_2$).

The "+3" ion, Co^{3+}, is the cobaltic ion. The cobaltic ion is not stable in aqueous solution but can exist in compounds like cobaltic fluoride (CoF_3) and cobaltic hydroxide ($Co[OH]_3$). The Co^{3+} ion forms a number of brightly colored complex compounds in which, most commonly, six other molecules or ions, called *ligands,* surround the cobaltic ion in an octahedral arrangement. Examples include reddish-violet chloropentammine cobaltichloride, $(Co[NH_3]_6Cl)Cl_2$; orange hexammine cobaltichloride, $Co(NH_3)_6Cl_3$; triammine cobaltinitrite, $Co(NH_3)_3(NO_2)_3$; aquapentammine cobaltihydroxide, $(Co[NH_3]_5[H_2O])(OH)_3$; aquapentammine cobaltichloride, $(Co[NH_3]_5H_2O)Cl_3$; and potassium cobaltinitrite, $K_3Co(NO_2)_6$, which is used as a paint pigment called cobalt yellow.

Occurring as it does in Group VIIIA, cobalt can have oxidation states even higher than "+2" or "+3". Because these higher oxidation states are very unstable and easy to reduce, compounds in which cobalt occurs in those high oxidation states are very powerful oxidizing agents. Examples include cobalt (IV) oxide, CoO_2; cobaltous cobaltite, $Co(CoO_3)$, in which the first cobalt atom is in the "+2" state and the second one is in the "+4" state; and tricobalt tetroxide, Co_3O_4, which can also be written as $2\ CoO \cdot CoO_2$, in which the first two cobalt atoms are in the "+2" state and the third atom is in the "+4" state.

COBALT: THE DISTINCTIVE BLUE

Cobalt blue, while expensive, is to many artists and glassmakers the quintessential blue pigment. Cobalt used in the coloration of glass and paint imparts a long-lasting, deep, bright blue that some label royal blue. This may be quite fitting, since some of the earliest cobalt-blue glass artifacts were probably made for decorating the abodes of pharaohs.

At least as far back as 1250 B.C.E., cobalt blue glass was being produced in Egypt. In 2005, archaeologists reported the discovery in Qantir, Egypt, of an ancient glass-manufacturing plant that produced cobalt blue *ingots*. The group also found high-quality, narrow-necked vessels as well as glass beads tinted to simulate the color of *lapis lazuli*. The production of ingots is an important clue to the commerce in the region. While it is unclear whether Egypt or Mesopotamia was first in glass production, trade in the substance went both ways, as illustrated in the map shown on page 82. The ingots could be shipped without fear of breakage to sites where they were melted down and shaped into new objects. A stock of cobalt blue ingots, found in a shipwreck dating from around the same century off the coast of Turkey, appears to have had its source in Amarna, Egypt, according to chemical analyses.

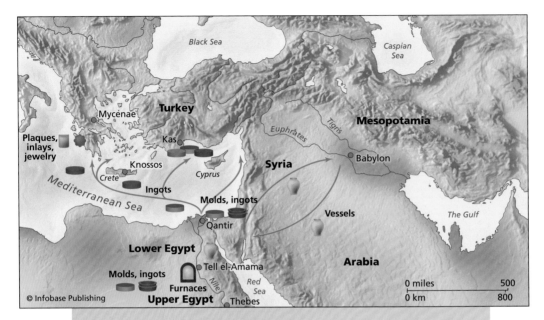

The ancient trade in glass, as shown on this map, involved transporting ingots of raw glass for processing into ornamental objects. *(Modeled after Preston Huey, Science)*

Another scientific capability, *X-ray fluorescence,* has identified cobalt in 14th–18th century ceramics from Valencia, Spain, although no record of cobalt mining in Spain before the 18th century has been found. The substance may have been shipped from mines in Germany or the Middle East.

Trade in cobalt for the purposes of art and ornamentation has been important for millennia. Perhaps the most famous and hallowed use of cobalt for beauty is in the cathedral at Chartres in France. The stained-glass windows, dating back to the 12th century, are renowned as the cobalt blue standard. Displaying an artisanship that probably cannot be found nowadays, the glass remains intact, unlike any other cathedral of the time. Again, however, there is no record of cobalt mining in the vicinity. In fact, scientific analysis has shown that this particular cobalt compound could not have been found in Europe, but is chemically similar to ancient Roman glass. The artisans probably used bits of old Roman glass to color their incomparable productions. How they obtained it is unknown, but the question is of great interest to some archaeologists.

Cobalt has been used in the coloration of glass for centuries.
(Johanne McCullough/Shutterstock)

THE CHEMISTRY OF NICKEL

Nickel is element 28, with a density of 8.9 g/cm^3, and ranks 23rd in order of abundance on Earth. Nickel-iron is a mineral in which nickel and iron both occur as the neutral metals. Nickel-iron is uncommon in terrestrial sources but fairly common in meteorites. It is believed that Earth's core is mostly nickel and iron. Together with iron and cobalt, nickel is one of the three magnetic metals.

Minerals that contain nickel include pentlandite, which is bronze in color and is a mixture of iron and nickel sulfides; garnierite, which is green in color and a mixture of nickel and magnesium silicates; and niccolite, which is copper-red in color and consists of nickel arsenide (NiAs). The primary source of nickel in North America is the mines in Sudbury, Ontario, Canada. Worldwide, the next most important source of nickel is New Caledonia.

Nickel's chemistry is very similar to that of cobalt. The nickelous ion (Ni^{2+}) is a bright green color in aqueous solution and is the only common nickel ion.

TECHNOLOGY AND CURRENT USES OF IRON, COBALT, AND NICKEL

Iron, mostly as steel, is one of the most important metals used by modern industry and technology. Steel, known for its strength and resistance to corrosion, is made by alloying iron with carbon and various metals. Its principal uses are in construction (e.g., buildings and bridges), tools, motor vehicles, weapons, and other structural applications. Stainless steel is an alloy of iron and chromium. Iron also has biophysical importance. Compounds of iron are added to plant fertilizers and used in nutritional supplements for people with iron deficiencies *(anemia)*.

Other uses of iron include powdered iron, used as a catalyst, and cast iron as an alloy of iron and carbon containing about 2.5 percent carbon. Iron is the principal component of permanent magnets. Various steel-based electrical-resistance alloys are used for electric heaters, clothes and hair dryers, stoves and ovens, and electrical heaters. It would be difficult to imagine modern civilization without extensive use of iron.

Second only to iron, cobalt is an important metal in permanent magnets. In addition, cobalt alloys are used in jet engines in the aerospace industry and in high-speed tools. Cobalt oxide is now the most used compound for giving glass, tiles, pottery, and ceramic glazes a distinctive blue color. Cobalt is also an element that is essential for life. Vitamin B-12 contains cobalt. Farmers sometimes add cobalt to salt blocks to prevent dietary deficiencies in their livestock. Various cobalt compounds are used as paint pigments, with potassium cobaltinitrite, $K_3Co(NO_2)_6$, a yellow pigment, being the most important. In modern nuclear medicine, the cobalt 60 isotope is an important artificially made radioactive isotope used in nuclear medicine. A source of gamma rays, Co-60 sterilizes medical instruments and is used to treat cancer.

Nickel is familiar to most people as the nickel coin, or 5-cent piece, currently minted in both the United States and Canada. Nickel has also replaced silver in dimes, quarters, half-dollars, and dollar coins. Nickel has several other uses in the metallic state. It is added to stainless steel to increase corrosion resistance, and copper-nickel alloys are used in water *desalination* plants. Batteries made of nickel and cadmium are rechargeable. A major use of nickel is in computer hard discs. In the aerospace industry, nickel cathodes are suitable for electroplating applications. In some applications, nickel is used in magnets.

Nickel compounds also have a number of applications. Nickel ammonium sulfate is employed in the electrolytic nickel-plating industry. Nickel compounds have applications in buildings, chemical manufacturing, communications, energy supply, environmental protection, food preparation, and water treatment. Finely divided nickel-based compounds serve as catalysts in a variety of applications, including the solidification of fats and oils.

THE PLATINUM METALS

The platinum metals are ruthenium (Ru), rhodium (Rh), palladium (Pd), osmium (Os), iridium (Ir), and platinum (Pt). These elements are among the rarest in Earth's crust. All of them are characterized by relatively high melting and boiling points and high densities. Palladium is 71st in abundance in Earth's crust with a density of 12.0 g/cm^3. Platinum is 72nd in abundance with a density of 21.5 g/cm^3. Osmium is 74th in abundance and is the densest element at 22.6 g/cm^3. Iridium is 77th, with a density of 22.4 g/cm^3. Ruthenium is 80th, with a density of 12.4 g/cm^3. Finally, rhodium is 81st, with a density of 12.4 g/cm^3. Almost the only elements that are rarer in Earth's crust are the noble gases and the radioactive elements that have to be produced artificially. Platinum is the most ductile metal in the group, meaning that it can be drawn into wires.

All of these metals share the common property of being relatively chemically inert. As a group, they are used heavily in industry as catalysts, with the most familiar application being the catalytic converters found in motor vehicles. Because of their scarcity but high demand in industry, these metals are traded in the commodity markets. Metals like platinum, palladium, and rhodium are often more precious than gold. In late 2009, when the spot price for gold was around $1,100 an ounce, the spot prices were $1,400 an ounce for platinum and $2,000 an ounce for rhodium. All of these prices fluctuate with demand, and in late 2009, the spot price for palladium was under $400 an ounce.

Historically, the countries that have been the major producers of platinum-group metals are Canada, Zimbabwe, South Africa, Russia, Australia, and the United States. Platinum can be found in nugget or

Platinum jewelry commands a high price. *(Ivan Stevanovic/ Shutterstock)*

small-grain form, in which its appearance is a silvery gray to white color. However, platinum is very rare. A typical ore might contain less than 0.1 ounce of platinum metals per ton of ore. The other metals in the group are often impurities in platinum deposits and may be present as alloys of platinum. A common platinum ore is sperrylite ($PtAs_2$). Otherwise, the other metals are mostly obtained as by-products in the processing of other metals such as nickel and copper.

THE ASTROPHYSICS OF THE PLATINUM METALS: Ru, Rh, Pd, Os, Ir, Pt

Evidence of early supernova activity in the outer reaches of the Milky Way is indicated by the detection of osmium and platinum in stars of the galactic halo. Iridium, platinum, palladium, rhodium, and the most abundant isotope of osmium, ^{192}Os, are all synthesized in supernovae via the rapid capture of a succession of neutrons by iron nuclei, which is called the *r-process*. Analysis of some meteorites, however, would seem to indicate that other osmium isotopes (^{184}Os, ^{186}Os, ^{188}Os, and ^{190}Os) are produced in massive stars via the slow or *s-process,* in which elements can build up over thousands of years in stellar atmospheres via neutron capture, with the requirement that iron 56 nuclei be present in the star in sufficient numbers to interact with free neutrons.

The s-process is also responsible for the existence of ruthenium 100. Ruthenium 99, however, is a radioactive fission element, resulting from the decay of technetium 99. Its presence has been detected in *stardust.* The formation process for the most abundant isotope of ruthenium (^{102}Ru) has not as yet been ascertained by astrophysicists.

Recent mass measurements of ruthenium ($^{90-92}$Ru), rhodium ($^{92-94}$Rh), and palladium ($^{94-95}$Pd) have led to the theory that these and neutron-deficient isotopes of other elements are most likely formed via proton capture. Supernova explosions produce a high density of photons that collide with some of the nuclei, breaking them up and freeing protons that are subsequently captured by other nuclei to make higher-Z elements.

(continued on page 90)

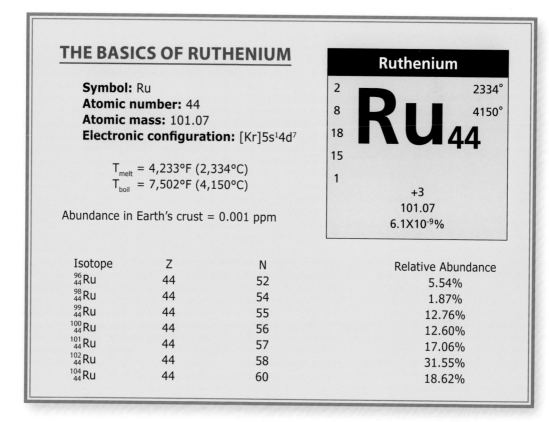

THE BASICS OF RUTHENIUM

Symbol: Ru
Atomic number: 44
Atomic mass: 101.07
Electronic configuration: [Kr]$5s^14d^7$

T_{melt} = 4,233°F (2,334°C)
T_{boil} = 7,502°F (4,150°C)

Abundance in Earth's crust = 0.001 ppm

Ruthenium

2
8
18
15
1

Ru$_{44}$

2334°
4150°

+3
101.07
6.1X10^{-9}%

Isotope	Z	N	Relative Abundance
$^{96}_{44}$Ru	44	52	5.54%
$^{98}_{44}$Ru	44	54	1.87%
$^{99}_{44}$Ru	44	55	12.76%
$^{100}_{44}$Ru	44	56	12.60%
$^{101}_{44}$Ru	44	57	17.06%
$^{102}_{44}$Ru	44	58	31.55%
$^{104}_{44}$Ru	44	60	18.62%

THE BASICS OF RHODIUM

Symbol: Rh
Atomic number: 45
Atomic mass: 102.9055
Electronic configuration: $[Kr]5s^14d^8$

T_{melt} = 3,567°F (1,964°C)
T_{boil} = 6,683°F (3,695°C)

Abundance in Earth's crust = 0.7 ppb

Isotope	Z	N
$^{103}_{45}$Rh	45	58

Rhodium

2		1964°
8	**Rh**$_{45}$	3695°
18		
16		
1		

+3
102.90550
1.12×10^{-9}%

Relative Abundance
100%

THE BASICS OF PALLADIUM

Symbol: Pd
Atomic number: 46
Atomic mass: 106.42
Electronic configuration: $[Kr]4d^{10}$

T_{melt} = 2,631°F (1,555°C)
T_{boil} = 5,365°F (2,963°C)

Abundance in Earth's crust = 0.6 ppm

Palladium

2		1554.9°
8	**Pd**$_{46}$	2963°
18		
18		
0		

+2 +4
106.42
4.5×10^{-9}%

Isotope	Z	N	Relative Abundance
$^{102}_{46}$Pd	46	56	1.02%
$^{104}_{46}$Pd	46	58	11.14%
$^{105}_{46}$Pd	46	59	22.33%
$^{106}_{46}$Pd	46	60	27.33%
$^{108}_{46}$Pd	46	62	26.46%
$^{110}_{46}$Pd	46	64	11.72%

THE BASICS OF OSMIUM

Symbol: Os
Atomic number: 76
Atomic mass: 190.23
Electronic configuration:
  [Xe]$6s^2 4f^{14} 5d^6$

T_{melt} = 5,491°F (3,033°C)
T_{boil} = 9,054°F (5,012°C)

Abundance in Earth's crust = 0.002 ppm

Osmium		
2		3033°
8	**Os**$_{76}$	5012°
18		
32		
14		
2	+3 +4	
	190.23	
	2.20X10^{-9}%	

Isotope	Z	N	Relative Abundance
$^{184}_{76}$Os	76	108	0.02%
$^{186}_{76}$Os	76	110	1.59%
$^{187}_{76}$Os	76	111	1.96%
$^{188}_{76}$Os	76	112	13.24%
$^{189}_{76}$Os	76	113	16.15%
$^{190}_{76}$Os	76	114	26.26%
$^{192}_{76}$Os	76	116	40.78%

THE BASICS OF IRIDIUM

Symbol: Ir
Atomic number: 77
Atomic mass: 192.217
Electronic configuration:
  [Xe]$6s^2 4f^{14} 5d^7$

T_{melt} = 4,435°F (2,446°C)
T_{boil} = 8,002°F (4,428°C)

Abundance in Earth's crust = 1 ppb

Iridium		
2		2446°
8	**Ir**$_{77}$	4428°
18		
32		
15		
2	+3 +4	
	192.217	
	2.16X10^{-9}%	

Isotope	Z	N	Relative Abundance
$^{191}_{77}$Ir	77	114	37.3%
$^{193}_{77}$Ir	77	116	62.7%

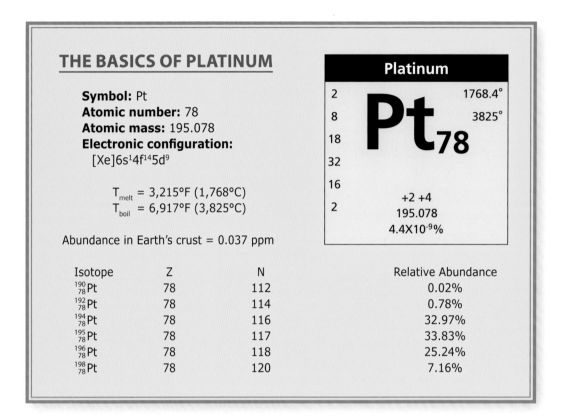

THE BASICS OF PLATINUM

Symbol: Pt
Atomic number: 78
Atomic mass: 195.078
Electronic configuration:
  $[Xe]6s^14f^{14}5d^9$

T_{melt} = 3,215°F (1,768°C)
T_{boil} = 6,917°F (3,825°C)

Abundance in Earth's crust = 0.037 ppm

Isotope	Z	N	Relative Abundance
$^{190}_{78}Pt$	78	112	0.02%
$^{192}_{78}Pt$	78	114	0.78%
$^{194}_{78}Pt$	78	116	32.97%
$^{195}_{78}Pt$	78	117	33.83%
$^{196}_{78}Pt$	78	118	25.24%
$^{198}_{78}Pt$	78	120	7.16%

Platinum

2		1768.4°
8	**Pt**$_{78}$	3825°
18		
32		
16		
2	+2 +4	
	195.078	
	$4.4X10^{-9}$%	

(continued from page 87)

DISCOVERY AND NAMING OF THE PLATINUM METALS

Of the elements comprising the so-called platinum metals, platinum itself was the earliest to be discovered. Like copper, silver, and gold, it is possible to find *native* (or pure) platinum metal. Because of its relative scarcity, platinum and its alloys had only minor use by people in the ancient world. It was in 1748, however, that platinum was brought to the attention of Europeans. Don Antonio de Ulloa (1716–95), a Spanish mathematician and naval officer, brought back samples of platinum from an expedition to Peru. Platinum was recognized as a noble metal with properties similar to those of gold and silver. De Ulloa believed that there were platinum mines in Peru. The word *platinum* comes from the Spanish word for silver—*platina*.

Platinum caused a sensation among chemists in Europe, many of whom wanted samples to study. Chemists showed that platinum could be made malleable by alloying it with silver and gold, that platinum dissolves in *aqua regia* (a combination of concentrated nitric and hydrochloric acids), and that under the right conditions it could be made to burn. Because of platinum's low chemical reactivity, however, few uses were found for it until the 20th century.

The next two platinum metals to be discovered were palladium and rhodium. Their discoverer was the English chemist and physicist William Hyde Wollaston (1766–1828). In 1803, Wollaston dissolved an impure sample of platinum in *aqua regia,* removed the excess acid, and precipitated a yellow solid by reaction with mercuric cyanide ($Hg[CN]_2$). Upon heating the precipitate, he obtained a small sample of a new metal that he named *palladium* in honor of the asteroid Pallas that also had recently been discovered. At first, chemists thought that palladium occurs only in association with platinum, but soon afterward, palladium was also discovered as an alloy with gold. Beginning in 1803, and finishing the work in 1804, Wollaston also succeeded in isolating rhodium from platinum ore. The name *rhodium* was derived from a Greek word meaning "rose-colored."

During the years 1803–04, Wollaston's mentor, Smithson Tennant (1761–1815), also worked on identifying the impurities in platinum ores. Tennant dissolved the ore in *aqua regia.* Upon neutralizing the resulting solution with base, he obtained both iridium and osmium. The name *iridium* was derived from the Greek word for "rainbow" because its salts have bright colors. The name *osmium* was derived from the Greek word for "odor" because osmium tetroxide (OsO_4) has a disagreeable odor.

The last platinum metal to be discovered was ruthenium. Ruthenium's discoverer was the Polish chemist Andrei Sniadecki (1768–1838). Working during the years 1807–09, Sniadecki obtained a new metal from a sample of platinum ore that he named *vestium* in honor of the asteroid Vesta that had just been discovered in 1807. Other chemists, however, were unable to duplicate Sniadecki's work, so his discovery remained in dispute. Although investigations after Sniadecki's death did, in fact, confirm his discovery, Sniadecki died before his claim was vindicated.

Definitive work on Sniadecki's new element was done later by Russian chemist Karl Karlovich Klaus (1796–1864). After studying platinum ores for several years, in 1844, Klaus validated Sniadecki's discovery but adopted the name *ruthenium* for the new metal in honor of his homeland, Russia.

THE CHEMISTRY OF THE PLATINUM METALS

Each of the platinum metals can exist in a variety of stable positive oxidation states as high as "+8". The most stable states are the following: Ru—2, 3, 4, 6, 7, 8; Rh—3, 4; Pd—2, 4; Os—3, 4, 6, 8; Ir—3, 4, 6; and Pt—2, 4. Each metal tends to react with oxygen gas (O_2) at elevated temperatures. Each metal also tends to react with chlorine gas (Cl_2) when heated.

The platinum metals tend to resist reaction with hydrochloric (HCl) or nitric (HNO_3) acids separately, with palladium being the only one that dissolves in nitric acid. They more readily dissolve—even if very slowly—in *aqua regia*. Even here, rhodium and iridium are exceptions and resist attack by *aqua regia*. Platinum metals tend not to form simple actions, but do form halide and oxide compounds.

More important than the chemical compounds they form are the catalytic properties of the platinum metals. As catalysts, they have high value in industry. Catalysts have the ability to speed up chemical reactions without themselves being consumed in the process. There are two general classes of catalysts. *Homogeneous catalysts* are present in the same *phase* as the chemicals whose reaction is being catalyzed. For example, inside the human body, important *proteins* called *enzymes* are present in the cell media and speed up the chemical reactions taking place in the cells that are necessary to support life processes. On the other hand, *heterogeneous* catalysts are present in a phase that is different than the phase the chemicals that are being catalyzed are in. In a chemical system, the platinum metals would be in the solid phase, whereas the reactions they catalyze are usually taking place in the gaseous or liquid phases.

TECHNOLOGY AND CURRENT USES OF THE PLATINUM METALS

The platinum metals are expensive. Therefore, they are usually used when no other elements would serve applications as well, thereby justifying their cost. A particularly important area of application is their

PALLADIUM AND COLD FUSION

In 1989, the University of Utah held a news conference in which two of its research scientists announced that they had discovered a way to produce fusion energy on a tabletop at room temperature. Dubbed "cold fusion," it stunned the scientific world because, if reproducible, the system could effectively provide the world's energy needs without the detrimental side effects of oil and coal burning. The scientists, Stanley Pons and Martin Fleischmann, had measured heat in a reaction that relied simply on palladium, *deuterium* (^{2}H), and electric current. The observed excess heat was given as evidence that more energy was being produced than was fed into the system in the form of electrical current—an overall gain in energy. It was also claimed that fusion by-products (neutrons and *tritium*) had been observed, although this was probably not the case. This news flash, however, was rapidly squelched when other scientists in the field found they could not reproduce the results of this simple-seeming experiment. The doubts seemed to be well founded, and skeptics started calling the work "junk science." Pons and Fleischmann were seen as foolish at best, and, at worst, charlatans.

Many respected scientists, however, decided to continue attempts to study the purported phenomenon, and over the years many have indeed observed excess heat from similar experiments. The basic setup involves embedding deuterium—hydrogen with a neutron connected to the proton—in a palladium metal lattice. This is done by immersing a palladium electrode in a *heavy water electrolyte* and passing current through it. Deuterium atoms seem to be absorbed into the lattice, producing energy under very particular conditions, which are not well understood, but the ratio of deuterium to palladium appears to be crucial, as does the purity level of the palladium. Researchers now have a higher success rate at heat production, and helium nuclei by-products have been identified in several cases, which would occur if two deuterium nuclei were to fuse, as in the following reaction.

$$^2_1H + {}^2_1H \rightarrow {}^4_2He + energy$$

(continues)

(continued)

While this normally should require extreme temperatures and pressures such as exist in the centers of stars, one hypothesis is that lattice vibrations within the palladium may provide the necessary force to effect the reaction.

Much work remains to be done, however, before the science is applicable. But, fortunately, many important agencies are showing interest. The Pentagon's Defense Advanced Research Projects Agency is funding experiments at the Naval Research Laboratory in Washington, D.C., and in August 2008, the American Chemical Society published the "Low Energy Nuclear Reactions Sourcebook" through Oxford University Press, a new bible of sorts for cold-fusion research. Despite his rough treatment in the past, in an April 2009 interview, Martin Fleischmann said, "The potential is exciting."

Mockup of test tube similar to that in which researchers Stanley Pons and Martin Fleischmann claim to have generated room-temperature fusion in 1989 *(George Frey/Time Life Pictures/Getty Images)*

use as catalysts. For example, platinum and rhodium are widely used in catalytic converters in motor vehicles. Platinum metal catalysts can also be used to remove trace impurities in other products. Other products for which their manufacturing processes are accelerated by platinum metal catalytic materials include gasoline, nitric and sulfuric acids, vitamins, antibiotics, hydrogen peroxide, and cortisone.

Some applications depend on the corrosion resistance and low chemical reactivity of platinum metals. For example, because of platinum's corrosion resistance, crucibles and electrodes may be made of platinum. Dental and medical instruments may contain platinum. Some spark plugs are now made of platinum. Palladium is used in the electrical industry and in current research on low-energy nuclear reactions.

Platinum is also used in jewelry and coinage. The coins are too expensive to circulate but are issued to collectors as investment items.

5

The Copper Group

The copper group consists of copper (Cu, element 29, *cuprum* in Latin), silver (Ag, element 47, *argentums* in Latin), and gold (Au, element 79, *aurum* in Latin). All were discovered and named in antiquity.

Copper, silver, and gold can all be found as the native metals, although usually they are found in ores. Copper is the principal metal in both bronze and brass; its early discovery permitted the fashioning of bronze implements, thus beginning the Bronze Age. Silver is rarer than copper and also less chemically reactive. The metal itself is used in jewelry, but compounds of silver are not because the compounds tend to be much less colorful. Gold is well known to most people, even if only by reputation. Because of its low chemical reactivity, gold is much more likely than either copper or silver to be found as the pure metal, usually in the form of nuggets or flakes.

THE BASICS OF COPPER

Element: Copper
Symbol: Cu
Atomic number: 29
Atomic mass: 63.546
Electronic configuration: $[Ar]4s^13d^{10}$

T_{melt} = 1,984°F (1,085°C)
T_{boil} = 4,644°F (2,562°C)

Abundance in Earth's crust = 68 ppm

Isotope	Z	N	Relative Abundance
$^{63}_{29}$Cu	29	34	69.17%
$^{65}_{29}$Cu	29	36	30.83%

Copper

2		1084.62°
8	**Cu$_{29}$**	2562°
18		
1		

+1 +2
63.546
1.70×10^{-6}%

THE BASICS OF SILVER

Element: Silver
Symbol: Ag
Atomic number: 47
Atomic mass: 107.868
Electronic configuration: $[Kr]5s^14d^{10}$

T_{melt} = 1,763°F (962°C)
T_{boil} = 3,924°F (2,162°C)

Abundance in Earth's crust = 0.08 ppm

Isotope	Z	N	Relative Abundance
$^{107}_{47}$Ag	47	60	51.84%
$^{109}_{47}$Ag	47	62	48.16%

Silver

2		961.78°
8	**Ag$_{47}$**	2162°
18		
18		
1		

+1
107.8682
1.58×10^{-9}%

Note: $^{94}_{47}$Ag has been synthesized at the National Superconducting Cyclotron Laboratory at Michigan State University. It is the heaviest odd N = Z isotope observed to date.

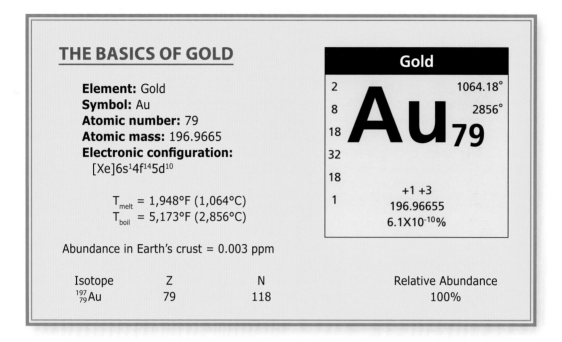

THE BASICS OF GOLD

Element: Gold
Symbol: Au
Atomic number: 79
Atomic mass: 196.9665
Electronic configuration:
  $[Xe]6s^14f^{14}5d^{10}$

T_{melt} = 1,948°F (1,064°C)
T_{boil} = 5,173°F (2,856°C)

Abundance in Earth's crust = 0.003 ppm

Isotope	Z	N	Relative Abundance
$^{197}_{79}$Au	79	118	100%

Gold

2
8
18
32
18
1

Au_{79}

1064.18°
2856°

+1 +3
196.96655
$6.1X10^{-10}$%

Copper is element number 29: It is a reddish metal with a density of 8.96 g/cm³. Copper is the 26th most abundant element in Earth's crust, and certainly one of the most familiar elements. Copper sometimes is found as the pure, or native, metal. All of copper's compounds and ions tend to be brightly colored—usually in hues of green, blue, or purple. Because of their colors, minerals containing copper, especially turquoise and malachite, are often used in jewelry. Copper is also an excellent conductor of electricity; one of its major uses is in electrical wiring. Its use in cookware is prevalent owing to its thermal conductivity.

Silver is element 47. The name itself is used to describe its color. Silver, at 10.5 g/cm³, is denser than copper. It is the 66th most abundant element in Earth's crust, and certainly one of the most familiar elements. Silver is an excellent conductor of electricity, better even than copper. Its cost, however, makes its use in ordinary electrical wiring impractical.

Gold is element 79: It is one of the densest of elements, 19.3 g/cm³, much denser even than lead. Gold ranks 73rd in abundance on Earth. Gold has a dull yellow appearance and is very resistant to corrosion or tarnish, much more so than copper or silver. Gold is the most malleable

Copper's thermal conductivity makes it useful in cookware.
(Ryan R. Fox/Shutterstock)

and ductile of all metals. Therefore, it finds important uses in industry and technology that are independent of its use in jewelry.

THE ASTROPHYSICS OF THE COPPER GROUP: Cu, Ag, Au

Silver and gold are synthesized in supernova explosions via the rapid capture of a succession of neutrons by iron nuclei, called the r-process. Until recently, astronomers believed that copper might be similarly synthesized, but it is now known that only 5–10 percent of copper in the universe comes from the r-process. A careful comparison of astrophysical spectra showing the copper-to-iron ratio in the Milky Way and the galaxy Omega Centauri indicates that the majority of copper nucleosynthesis occurs in what is termed the *weak s-process,* which can only occur in stars that are at least eight times as massive as the Sun.

When such stars begin to fuse helium in the core, they enter the red-giant stage and produce carbon by fusion of helium nuclei. Once carbon is formed, nitrogen can form in the following steps:

$$^{12}_{6}\text{C} + ^{1}_{1}\text{H} \rightarrow ^{13}_{7}\text{N} + \text{energy}$$

$$^{13}_{7}\text{N} \rightarrow ^{13}_{6}\text{C} + e^{+} + ^{1}_{0}\nu$$

$$^{13}_{6}\text{C} + ^{1}_{1}\text{H} \rightarrow ^{14}_{7}\text{N} + \text{energy}.$$

Nitrogen 14 can then capture two alpha particles to form ^{22}Ne. Neon 22 capturing an alpha particle produces the neutrons needed in the weak s-process by the following reaction.

$$^{22}_{10}\text{Ne} + ^{4}_{2}\alpha \rightarrow ^{25}_{12}\text{Mg} + ^{1}_{0}\text{n}$$

Over time, iron nuclei, gravitationally collected from the immediate vicinity during the star's formation, absorb neutrons produced in the above reaction, leading to the synthesis of copper.

A HISTORY OF COINAGE

Copper was the first metal to be used in coins. Unfortunately, copper weathers fairly easily, so relatively few ancient copper coins have survived in pristine condition. Even more modern copper coins—such as from the American Colonial and early National Periods—tend to be rare in mint condition. Reflecting copper's lower monetary value than silver or gold, the most common copper coin is the cent, or penny.

Silver has been used in coins since ancient times. Because silver tends to react less readily with environmental chemicals than copper does, silver coins tend to survive in more intact condition. In fact, some of the best examples of ancient coins tend to be ones made of silver. Because silver is more valuable than copper is, for many years U.S. dimes, quarters, half dollars, and dollars historically were minted from silver.

Gold coins are prized among collectors. Even ancient coins can be found in relatively pristine condition.

The American colonies mainly relied on coinage from Europe and Latin America. (One of the first tasks of the newly established United States after the Revolutionary War was to establish a United States mint.) The term *dollar* was derived from the German *thaler*. Because most coins were large and high in value, it was common practice to cut

Silver and gold have been used in coinage for millennia. *(Denis Miraniuk/Shutterstock)*

them into pieces to make small change. Most British coins were made of copper. Another common coin was the Spanish milled dollar, a silver coin minted throughout Latin America from 1732 to 1772. Spanish dollars had a value of 8 *reales* and were also known as pieces of eight—a term familiar from pirate lore.

Individual colonies did produce a few rudimentary coins similar to British small coins of the time—penny, twopence, threepence, shilling, and farthing, for example. These coins were usually made of copper. The first coin issued by the United States was the *fugio* cent, a copper coin first made in 1787. Regular issue of United States coins began in the 1790s with a copper half-cent piece and a copper large cent similar in size to the current quarter. The first silver coins, also issued in the 1790s, included a small half dime, a dime, a quarter, a half dollar, and a dollar. The half-cent and large cent coins were discontinued after 1857.

Beginning in 1856, the flying eagle small cent (the size of the current Lincoln cent) was issued. The mint experimented with a copper two-cent piece from 1864 to 1873, when it was discontinued. Similarly, the mint tried a small silver three-cent piece (called a trieme) from 1851 to 1873 and then discontinued it. (The coin was smaller in size than a cent and easily lost.) During this time, nickel obtained from the Gap Mine in Lancaster County, Pennsylvania, was used in U.S. coins. A nickel three-cent piece was introduced in 1865 but discontinued after 1889. The first so-called nickel five-cent piece—called the shield nickel—was issued in 1866. Although a five-cent piece is called a nickel, its composition is actually 75 percent copper and only 25 percent nickel, the latter mostly as a coating over a copper interior. The Gap Mine continued to be the nation's source of nickel until 1891, when nickel was obtained more productively from mines in Sudbury, Ontario, Canada.

Dimes and quarters were 90 percent silver and 10 percent copper until 1964. Because of the rising price of silver, the issuance of silver coins was discontinued in the United States. Beginning in 1965, dimes and quarters have had a composition that is 75 percent copper and 25 percent nickel. Half dollars were 90 percent silver and 10 percent copper through 1964. From 1965 to 1970, the composition was 80 percent silver and 20 percent copper. In 1971, silver half dollars were replaced with coins having a core of pure copper surrounded by a layer consisting of 75 percent copper and 25 percent nickel.

Beginning in 1794, early U.S. dollars were 89 percent silver and 11 percent copper. The minting of dollar coins was halted after 1804 and not resumed until 1836. Dollars consisted of 90 percent silver and 10 percent copper until 1935, when minting of dollar coins ceased. Eisenhower dollars began to be issued in 1971 with different compositions, depending on the mint. Coins minted in Philadelphia and Denver were 75 percent copper and 25 percent nickel. Coins minted in San Francisco had an outer layer that was 80 percent silver and 20 percent copper with an inner core consisting of 21 percent silver and 79 percent copper. Silver dollars were discontinued after 1976. The Susan B. Anthony dollars, issued in 1979–81 and again in 1999, consisted of a pure copper inner core and an outer core of 75 percent copper and 25 percent nickel. The Sacagawea dollar, the minting of which began in 2000, has an outer layer of manganese brass with a composition that is 77 percent copper, 12 percent zinc, 7 percent manganese, and 4 percent nickel, with a pure copper inner core. The coins in the presidential dollar series that began in 2007 have the same composition as the Sacagawea coins.

The first gold coins minted in the United States were issued in 1795. For more than a century, gold coins intended for general circulation had the following denominations: $1, $2.50, $3, $4, $5, $10, and $20. The $10 coin was called an eagle. Therefore, the $2.50 coin was a quarter eagle, the $5 coin a half eagle, and the $20 coin a double eagle. The composition of gold coins was generally 90 percent gold and 10 percent silver. The production of gold coins was discontinued after 1933. All gold was removed from circulation, and a large quantity of coins was melted down.

In addition to coins minted for general circulation, the United States mint has also produced many coins as commemorative pieces or as *bullion*. Today, these coins can be made of silver, gold, or platinum, and are usually only purchased by collectors or investors.

Canada has a history of coinage that parallels that of the United States. One distinctive difference between the coins of the two countries is that, by law, coins minted by the United States government cannot bear the likeness of a living person. (This tradition dates back to George Washington, who refused to allow his likeness to appear on the first U.S. coins. He did not want people to compare him to the king of England.) On the other hand, because Canada belongs to the British Commonwealth of Nations, Canadian coins have always displayed the likeness of Britain's reigning monarch on its coins. Canada became a Confederation in 1867, although Canada began issuing its own coins in 1858. Beginning with the likeness of Queen Victoria appearing on coins, the tradition continues with the likeness of Queen Elizabeth II adorning current coins.

THE CHEMISTRY OF COPPER

Copper is an essential element for humans and for plant growth. In certain mollusks and crustaceans, copper in hemocyanin substitutes for the iron found in the hemoglobin of most animals as the carrier of oxygen through the bloodstream. In most animals, iron imparts a red color to blood. In mollusks and crustaceans that instead use hemocyanin, the copper imparts a blue color to blood.

Copper is one of the few metals that can be found in native form. For example, deposits of pure copper metal occur in the Upper Peninsula of

Michigan. Pure copper metal is red—a color familiar because of copper coins, especially the Lincoln cent in the United States.

Mostly, however, copper is found in ores. About 20 percent of the world's copper comes from the United States. The largest mines in the United States are in Arizona, Utah, and Montana, with smaller mines in Michigan, Nevada, and New Mexico. Arizona leads the United States in copper production, with about 1.6 billion pounds of copper being recovered every year at a market value of $5.5 billion. Worldwide, there are mines on every continent except Antarctica. In recent years, Chile has developed extensive copper mines and become a leading producer of copper. Copper ores are easily reduced to copper metal by electrolysis, a process called *electrowinning*.

A distinguishing feature of copper is that—unlike other familiar metals like aluminum, zinc, tin, lead, nickel, magnesium, chromium, and manganese—it does not react with aqueous solutions of hydrochloric or sulfuric acids. Because of copper's lesser tendency to be oxidized, more powerful oxidizing acids are required. Metallic copper is usually dissolved in concentrated nitric acid, which results in the rapid evolution of the reddish-brown, toxic gas nitrogen dioxide (NO_2), as shown in the following chemical equation:

$$Cu\ (s) + 4\ HNO_3\ (aq) \rightarrow Cu(NO_3)_2\ (aq) + 2\ NO_2\ (g) + 2\ H_2O\ (l).$$

In the "+1" oxidation state, copper exists as the colorless cuprous ion (Cu^+). In the "+2" state, it is found as the sky-blue cupric ion (Cu^{2+}). The cuprous ion is unstable, and tends to disproportionate to copper metal and the cupric ion, as shown by the following equation:

$$2\ Cu^+\ (aq) \rightarrow Cu\ (s) + Cu^{2+}\ (aq).$$

Cu^+ does, however, form a few compounds such as cuprous oxide (Cu_2O). Cu_2O is red in color, is found in nature, and is used to color porcelain. Cu_2O dissolves in ammonia, cyanide solutions, and hydrochloric acid to give the complex ions $Cu(NH_3)_2^+$, $Cu(CN)_3^{2-}$, and $CuCl_2^-$, respectively. Copper in the "+1" state bonds more readily to carbon than it does in the "+2" state. Thus, Cu^+ has an extensive chemistry with organic compounds.

In aqueous solution, the cupric ion tends to be surrounded by water molecules, forming the $Cu(H_2O)_4^{2+}$ complex ion. All Cu^{2+} compounds or ions are colored—most commonly as shades of blue, green, or black. In the "+2" state, copper forms a large number of compounds, including black cupric oxide (CuO), green cupric hydroxide ($Cu[OH]_2$), green cupric carbonate ($CuCO_3$), black cupric sulfide (CuS), green cupric chloride ($CuCl_2$), blue cupric sulfate ($CuSO_4$), blue cupric nitrate ($Cu[NO_3]_2$), blue cupric phosphate ($Cu_3[PO_4]_2$), and reddish-brown cupric ferrocyanide ($Cu_2Fe[CN]_6$). The nitrate and chloride are soluble in water. Otherwise, most cupric compounds are insoluble.

THE CHEMISTRY OF SILVER

Until the beginning of the 20th century, silver was used mostly for coinage. For more than a century, however, silver has been important in industry, especially in photographic film. As early as 1727, experiments were being done to make images using compounds of silver. Silver chloride ($AgCl$) darkens upon exposure to light. Upon treatment of the film, a *negative image* develops, from which prints may be made.

In analytical tests, chemists make use of the fact that relatively few metal ions—namely Ag^+, Pb^{2+}, and Hg_2^{2+}—precipitate with halide ions (Cl^-, Br^-, I^-). Therefore, if mixtures of metal ions in solution are treated with dilute hydrochloric acid (HCl), only $AgCl$, $PbCl_2$, and Hg_2Cl_2 will precipitate. All three compounds are white in color. $AgCl$ may be separated from the other two solids by the addition of dilute aqueous ammonia (NH_3), since $AgCl$ dissolves in the presence of ammonia, as shown by the following equation:

$$AgCl\ (s) + 2\ NH_3\ (aq) \rightarrow Ag(NH_3)_2^+\ (aq) + Cl^-\ (aq).$$

$PbCl_2$ reacts with aqueous ammonia to form $Pb(OH)_2$, which is also white, so evidence of reaction is not obvious. Hg_2Cl_2 reacts with aqueous ammonia to form a mixture of metallic mercury and mercury ammono chloride ($HgNH_2Cl$). The silver complex ion can be separated from the mixture by *filtration*.

Silver undergoes similar reactions with the Br^- and I^- ions. The solubility of each successive silver halide salt decreases with increasing atomic weight of the halide. Thus, $AgCl$ is the most soluble of the three compounds, AgBr is less soluble, and AgI is the least soluble of all three. The colors also differ; AgBr and AgI are yellow in color. In addition, AgBr and AgI are less reactive with aqueous ammonia. Whereas AgCl dissolves in a dilute solution of ammonia, AgBr requires concentrated ammonia to dissolve, and AgI does not dissolve at all in ammonia.

The silver ion (Ag^+) forms insoluble compounds with most of the common negative ions. In approximate order of decreasing solubility, examples include the following: silver carbonate (Ag_2CO_3), silver oxalate ($Ag_2C_2O_4$), silver chromate (Ag_2CrO_4), silver oxide (Ag_2O), the silver halides (AgCl, AgBr, and AgI), silver cyanide (AgCN), and silver sulfide (Ag_2S). Most of these compounds are white in color, with the darkest compounds being reddish-brown silver chromate and black silver sulfide. Silver sulfide is the familiar tarnish on silverware that results from the following reaction between silver metal and small amounts of hydrogen sulfide gas (H_2S) in air:

$$Ag \text{ (s)} + H_2S \text{ (g)} \rightarrow Ag_2S \text{ (s)} + H_2 \text{ (g)}.$$

In high school and college laboratories, the most common silver reagent is silver nitrate ($AgNO_3$). Silver nitrate is also used as an *antiseptic*.

THE CHEMISTRY OF GOLD

Gold is found in nature mostly as the native metal, often in conjunction with native silver and platinum metals. Two examples of gold-containing compounds are tellurides—$AuTe_2$ and a mixture of gold and silver, $AuAgTe_4$. Seawater contains minute quantities of gold (about 0.1 to 0.2 milligram of gold per ton of seawater) washed into streams and eventually into the oceans. Compounds of gold contain gold in one of two oxidation states: "+1" or "+3". The "+1" ion, Au^+, is called the aurous ion; the "+3" ion, Au^{3+}, is called the auric ion.

Beginning in about 500 B.C.E., gold was recovered from ore by *amalgamating* the gold with liquid mercury. After the gold had been separated from the ore, the mercury could be removed by heating. In about 1890, the modern process of recovering gold with cyanide treatment greatly boosted the ability to extract gold from small deposits. In this process, cyanide combines with gold by the following reaction:

$$4 \text{ Au (s)} + 8 \text{ CN}^- \text{ (aq)} + O_2 \text{ (g)} + 2 \text{ H}_2O \text{ } (l) \rightarrow$$
$$4 \text{ Au(CN)}_2^- \text{ (aq)} + 4 \text{ OH}^- \text{ (aq)}.$$

The reaction is very slow and requires several days to be accomplished. Subsequent reaction with zinc metal reduces the gold back to the neutral metal.

One of the fascinations ancient and medieval alchemists had with gold was gold's very low chemical reactivity. Around the beginning of the 14th century, alchemy took a giant leap forward with the discoveries of the strong mineral acids—hydrochloric (HCl), nitric (HNO$_3$), and sulfuric (H$_2$SO$_4$). Making good use of such powerful reagents, alchemists found that nearly all metals were soluble in either HCl or H$_2$SO$_4$,

THE RUSH FOR GOLD

Gold has probably been the most sought-after element in all of history. Prized for jewelry, coinage, and a variety of ornamental uses, ships have been launched, wars have been fought, and men and women have been murdered because of gold. Medieval alchemists spent many years searching for the *philosopher's stone* that would turn common metals like lead into gold. The Spaniards scoured the Western Hemisphere in their quest for the gold of the Inca and Aztec Indians, and for the Seven Cities of Cibola (the "Cities of Gold"). Famous gold rushes have taken place throughout western North America, including California, Colorado, and the Yukon. Schoolchildren learn in their history classes about the "49-ers," the prospectors who sailed around Cape Horn, crossed the Isthmus of Panama, or crossed the American continent by wagon train, lured by the California mother lode. Had it not been for the California gold rush, the admission of California to the Union would undoubtedly have taken place many years later than it did.

or possibly in HNO_3—but that gold did not dissolve. Only when concentrated HCl and HNO_3 were combined to make a new reagent, *aqua regia*, could gold be dissolved in the following reaction:

$$Au\ (s) + 5\ H^+\ (aq) + 4\ Cl^-\ (aq) + NO_3^-\ (aq) \rightarrow$$
$$HAuCl_4\ (aq) + NO\ (g) + 2\ H_2O\ (l).$$

Even the name *aqua regia* reflects its use. Since gold was primarily the property of the king, it made sense that the "king's water" was required to dissolve it.

The auric ion (Au^{3+}) is much more stable than the aurous ion (Au^+). In aqueous solution, only aurous cyanide $[Au(CN)_2^-]$ is more stable than the auric form $[Au(CN)_4^-]$. In the absence of moisture, auric chloride decomposes when heated to form aurous chloride, as shown in the following equation:

$$AuCl_3\ (g) \rightarrow AuCl\ (s) + Cl_2\ (g).$$

There does exist an Au^{2+} ion, but it is less stable than either Au^+ or Au^{3+}.

HARD-ROCK MINING: THE ENVIRONMENTAL COSTS

Nearly all gold, silver, and copper mines presently active in the United States are open-pit or strip mines. Heavy machinery is required to dig, remove, and break rock to get to the precious-metal veins or to carry ore that may have even a small gold or silver content to heaps or vats where it can be further refined for purity. The average size of such a project in the United States is 1,000 acres, each displacing all *flora* and *fauna* in the immediate area. The process of soil and bedrock removal generally exposes previously buried sulfur-containing minerals to air and weather. Rainfall or immersion of these minerals in streams produces sulfuric acid, which can dissolve heavy metals that are normally bound in soils. Toxic metals like mercury, arsenic, cadmium, chromium, and lead thereby become incorporated in streams or groundwater. This *acid mine drainage* also has the effect of changing the pH of the water, which can be extremely detrimental to aquatic life.

Until around 1970, mercury was used in hydraulic mines that incorporated water in sluices to recover small bits of gold from gravel

Workers labor inside a silver mine. *(Jduggan/Dreamstime.com)*

and sediment. Gold and mercury have a tendency to amalgamate and sink, while lighter bits of dirt and sand are washed away by the water poured into the sluice, after which the gold can be separated from the mercury. Naturally, a good portion of mercury was also washed away:

The U.S. Geological Survey estimates that 3 million pounds of mercury have been lost over the history of its use in hard-rock mines. Environmental effects are most problematic in groundwater contamination, inhalation, and improper handling of old abandoned equipment. (See "Environmental Mercury Risks" in chapter 6.)

From 1970 to the present, the preferred method to separate gold and silver from crushed ore in hard-rock mines has been *cyanide leaching*. Cyanide has a propensity to bond with gold or silver to form a water-soluble compound. Heaps or vats of crushed rock are sprayed with cyanide solution that then seeps through the material and is collected below. In theory, all the cyanide solution is then collected, separated, and recycled. In practice, however, spills and intentional discharges have led to serious conditions, with complete ecosystems destroyed and drinking water supplies contaminated.

The waste factor must also be considered. Every ton of pure copper extracted produces about 100 tons of waste. For gold, the waste-to-value ratio is on average 300 to 1, and for gold, about 80 percent of production is to supply jewelers' needs. Perhaps the most profligate example of destruction is the gold, silver, and copper mine in Mount Ertsberg, New Guinea. The owners estimate that the mine generates 700,000 tons (635,029 tonnes) of waste daily. Environmental studies show that, in 90 square miles (233.1 km²) of downstream wetlands, all fish life has vanished.

Regulatory language against excessive cyanide release, whether accidental or intentional, exists, but enforcement is effectively nonexistent. Lawsuits have been successfully filed, but the results have turned out to cost the offending companies very little. The continuing demand for the copper group metals may, in the future, bring this issue to the attention of the public, for better or worse.

TECHNOLOGY AND CURRENT USES OF COPPER

Since ancient times, copper has been alloyed with tin to make bronze, and alloyed with zinc to make brass. Copper and its alloys are used in pipes and tubes in industry, in commercial buildings, and in homes. Copper has also been used to make coins since the beginning of coinage.

Copper cookware is used extensively in food processing. Because of copper's high *electrical conductivity* (second only to silver), copper is used extensively in electrical wiring.

A number of copper compounds have useful applications. Cupric acetate ($Cu[C_2H_3O_2]_2$) is used as a *fungicide* and *insecticide*. Cupric hydroxide ($Cu[OH]_2$) and basic cupric carbonate, which has the same composition as the mineral malachite ($CuCO_3 \cdot Cu[OH]_2$), are paint pigments. Cuprous cyanide ($Cu_2[CN]_2$) and cuprous iodide (Cu_2I_2) are used in organic syntheses. Cuprous oxide (Cu_2O) is a fungicide. Cupric oxide (CuO) is used to color ceramics. Cupric sulfate ($CuSO_4$) is used in copper plating, water treatment, and coating steel. Cupric sulfate is also a soil-amendment product, a pesticide, and a *germicide*. Cuprous chloride ($CuCl$) is a catalyst for various organic reactions. Cupric chloride ($CuCl_2$) is a catalyst in several organic syntheses. Cupric chloride is also a wood preservative and disinfectant.

TECHNOLOGY AND CURRENT USES OF SILVER

Silver coinage has been used since ancient times. Silver or alloys of silver are important components of some batteries, electrical contacts, and printed circuits. Sterling silver is used for tableware and jewelry. Some dentists use an *amalgam* of silver and other metals mixed with mercury to fill cavities in teeth, though this practice is discouraged by the World Health Organization due to the detrimental effects of mercury on the human system. Silver sometimes is used for the reflective backing of mirrors.

Silver salts have been extremely important to the film and photography industry. Silver salts and oxides are important catalysts in organic syntheses. Silver iodide (AgI) is used in *cloud seeding*. Silver nitrate is used as an antiseptic.

TECHNOLOGY AND CURRENT USES OF GOLD

Most of the uses of gold are of the metal itself, not of its compounds. Uses of the metal are manifold. Nearly all gold, 78 percent, is made into jewelry and ornamentation. Gold leaf and decorative gold have been used for centuries. Only about 10 percent of the world's supply of gold is used in finance, including gold bullion, coins, and medals.

The remaining 12 percent of all gold is used in medical and dental electronics applications and in dental restorations. Medical applications of gold include anticancer treatments, arthritis treatments, implants, and pregnancy tests. Gold flakes and *nanoparticles* are used in the cosmetics industry.

The construction of the U.S. *Columbia* space shuttle utilized gold brazing alloys, gold plastic film coatings, and gold electrical contacts. A thin, transparent layer of gold coats the face masks of space helmets to reflect heat and light. In the same manner, gold coatings protect the Hubble Space Telescope from corrosion.

Gold is used in computer components. The automobile industry is beginning to use gold in catalytic converters. In addition, the use of gold catalysts in fuel cells is being investigated. Gold bonding wire and gold electroplating are used in electronics and electrical contacts. Research into uses for gold in nanotechnology applications is ongoing.

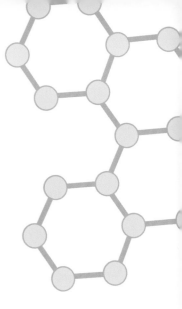

6

The Zinc Group

The zinc group consists of zinc (Zn, element 30), cadmium (Cd, element 48), and mercury (Hg, element 80, *hydragyrum* in Greek). Zinc, cadmium, and mercury are the only naturally occurring elements in Group IIB and occupy the last column of the *d* block of elements. Various authors classify these three elements differently. Authors who adhere strictly to the definition of a transition metal as an element with an only partially occupied *d* subshell will exclude zinc, cadmium, and mercury because their *d* subshells are filled. That is, each element has an electronic configuration in which the outermost subshell configuration is nd^{10}, where $n = 3$ for zinc, 4 for cadmium, and 5 for mercury. A *d* subshell can hold at most 10 electrons, so that, in each case, the subshell is filled. Other authors include zinc, cadmium, and mercury with the transition metals because their properties are more similar to the transition metals than to the post-transition metals, whose properties are determined by partially filled p subshells. (In this multivolume set devoted to the periodic table, it was decided to include zinc, cadmium, and mercury with the transition metals.)

Zinc is element 30 with a density of 7.1 g/cm³. It is 24th in abundance in Earth's crust. Several ores contain zinc, including zinc sulfide (ZnS) in the mineral sphalerite (also known as zinc blende), zinc oxide (ZnO) in the mineral zincite, franklinite (a mixture of zinc, manganese, and iron oxides), zinc silicate (Zn_2SiO_4) in the mineral willemite, and the mineral calamine ($Zn_2[OH]_2SiO_3$). Metallic zinc can be recovered by reduction with carbon. Usually, however, zinc is produced through electrolytic processes. A fresh zinc surface has a shiny metallic luster. However, it tarnishes easily. Ordinarily, zinc is hard and brittle, but at higher temperatures (more than 100°C) it becomes malleable. In the case of recovery of zinc from zinc sulfide, sulfur dioxide (SO_2) is a by-product. Most of the sulfur dioxide is used to produce sulfuric acid.

THE BASICS OF ZINC

Element: Zinc
Symbol: Zn
Atomic number: 30
Atomic mass: 65.39
Electronic configuration: [Ar]$4s^2 3d^{10}$

T_{melt} = 787°F (420°C)
T_{boil} = 1,665°F (907°C)

Abundance in Earth's crust = 79 ppm

Zinc			
2			419.53°
8	**Zn**		907°
18		**30**	
2			
	+2		
	65.39		
	4.11X10⁻⁶%		

Isotope	Z	N	Relative Abundance
$^{64}_{30}$Zn	30	34	48.63%
$^{66}_{30}$Zn	30	36	27.90%
$^{67}_{30}$Zn	30	37	4.10%
$^{68}_{30}$Zn	30	38	18.75%
$^{70}_{30}$Zn	30	40	0.62%

THE BASICS OF CADMIUM

Element: Cadmium
Symbol: Cd
Atomic number: 48
Atomic mass: 112.411
Electronic configuration: $[Kr]5s^24d^{10}$

T_{melt} = 610°F (321°C)
T_{boil} = 1,413°F (767°C)

Abundance in Earth's crust = 0.15 ppm

Cadmium		
2		321.07°
8	**Cd**48	767°
18		
18		
2		
	+2	
	112.411	
	5.3X10⁻⁹%	

Isotope	Z	N	Relative Abundance
$^{106}_{48}Cd$	48	58	1.25%
$^{108}_{48}Cd$	48	60	0.89%
$^{110}_{48}Cd$	48	62	12.49%
$^{111}_{48}Cd$	48	63	12.80%
$^{112}_{48}Cd$	48	64	24.13%
$^{113}_{48}Cd$	48	65	12.22%
$^{114}_{48}Cd$	48	66	28.73%
$^{116}_{48}Cd$	48	68	7.49%

THE BASICS OF MERCURY

Element: Mercury
Symbol: Hg
Atomic number: 80
Atomic mass: 200.59
Electronic configuration:
   $[Xe]6s^24f^{14}5d^{10}$

T_{melt} = -38°F (-39°C)
T_{boil} = 674°F (357°C)

Abundance in Earth's crust = 0.067 ppm

Mercury		
2		-38.83°
8	**Hg**80	356.73°
18		1477°
32		
18		
2	+1 +2	
	200.59	
	1.11X10⁻⁹%	

Isotope	Z	N	Relative Abundance
$^{196}_{80}Hg$	80	116	0.15%
$^{198}_{80}Hg$	80	118	9.97%
$^{199}_{80}Hg$	80	119	16.87%
$^{200}_{80}Hg$	80	120	23.10%
$^{201}_{80}Hg$	80	121	13.18%
$^{202}_{80}Hg$	80	122	29.86%
$^{204}_{80}Hg$	80	124	6.87%

The Eagle River threads its way through a zinc-mining area in Eagle County, Colorado. *(U.S. Geological Survey)*

Cadmium is element 48: It has a density of 8.65 g/cm^3 and ranks 65th in abundance in Earth's crust. Pure cadmium metal is white with a slight bluish tint. Cadmium is a relatively soft metal, making it fairly ductile and malleable at room temperature. Although it finds uses in technological applications, there are a number of health concerns about cadmium's toxicity. Cadmium is also *carcinogenic* and *teratogenic,* meaning that it causes both cancer and birth defects. These concerns can pose health hazards to persons whose job is to recycle materials that may contain cadmium. Most cadmium is extracted from zinc ores in which the cadmium content is usually about 0.5 percent the zinc content. During the reduction process, metallic cadmium is obtained first, followed by metallic zinc. The only important cadmium ore is greenockite, which is yellow in color and is composed of cadmium sulfide (CdS).

Mercury is element 80: It is rather dense for a liquid, having a density of 13.5 g/cm^3 at room temperature. The principal ore from which mercury is obtained is reddish cinnabar (HgS). Mercury has been found in nature as the pure metal and also in conjunction with silver and gold. The element was known to ancient people. They were fascinated by its

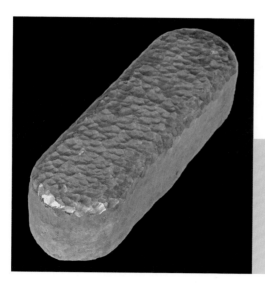

A museum-grade ingot sample of cadmium. This soft, bluish-white metallic element occurs in trace amounts in zinc, copper, and lead ores. *(Theodore Gray/Visuals Unlimited)*

liquid property under normal conditions, yet it has a silvery color much like other metals that were known at the time. Through medieval times, alchemists and medical practitioners called mercury *quicksilver* in reference to its liquid properties.

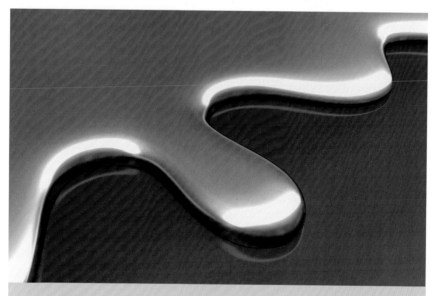

Through medieval times, alchemists and medical practitioners called mercury *quicksilver* in reference to its liquid properties. *(iDesign/Shutterstock)*

THE ASTROPHYSICS OF THE ZINC GROUP: Zn, Cd, Hg

Zinc, cadmium, and mercury are a curious group as regards their modes of production in stars. As explained in chapter 4, nickel 56 is created in Type I supernovae, when, at the instant of the explosion, iron 52 nuclei that had been present in the star collide with helium nuclei (alpha particles). The nickel thus formed could also capture helium nuclei, synthesizing zinc 60 in the following reaction.

$$^{56}_{28}\text{Ni} + ^{4}_{2}\alpha \rightarrow ^{60}_{30}\text{Zn}$$

In this sense, zinc can be classified as an alpha-process element.

A recent comparison of some Type II supernovae (see chapter 4) with galactic halo stars indicates that another mechanism must also be responsible for some stellar zinc production. While the amount synthesized is sensitive to mass mixing and explosion energy, which is different for every supernova, it was found that the majority of iron in the universe forms via the r-process, which is the rapid capture of a succession of neutrons by iron nuclei during Type II supernova events.

Studies of interstellar cadmium, on the other hand, show that this element is most likely formed over thousands of years in helium-fusing shells of asymptotic giant branch stars via neutron capture. Because this synthesis proceeds relatively slowly due to a low density of neutrons, it is called the s-process (sometimes called the main s-process in order to distinguish it from the weak s-process, described in chapter 5).

Some of the most puzzling astrophysical observations concern the detection of excess mercury in chemically peculiar (CP) stars. Because mercury is so heavy, it ought to sink beneath the surface of its host star, in which case its spectrum would be effectively blocked by the emissions of lighter elements on the surface. Yet a certain type of CP star (labeled HgMn stars for their overabundant elements) show excesses of mercury that are orders of magnitude higher than the content in the Sun. Even more curious is evidence of an uneven distribution of mercury in some stars, leading astrophysicists to hypothesize about possible mercury cloud formation. Strange magnetic field distributions are involved along with an unknown mechanism for separating isotopes within these stars. Because the physics of CP stars is so complex, this area of research has fostered the development of novel instrumentation and modeling techniques, and will probably continue to do so for years to come.

DISCOVERY AND NAMING OF ZINC, CADMIUM, AND MERCURY

Compounds of zinc have been used since ancient times. Ancient Greek writings refer to the application of zinc ointments to heal wounds. Zinc was added to copper to make brass. In China and India, metallic zinc was reported to have been smelted from its ores as early as the 13th century C.E. After 1700, zinc was imported to Europe regularly from these Asian countries. It was not until the 1700s, however, that zinc was produced in Europe. The temperature required to reduce the zinc in ores to zinc metal is higher than the boiling point of zinc. Therefore, if the reduction process was not done correctly, it was common for the zinc to vaporize before the metal could be obtained.

In 1746, the German chemist Andreas Sigismund Marggraf (1709–82) succeeded in obtaining zinc by the reduction of the mineral calamine (zinc oxide, ZnO). In the 19th century, zinc was discovered in several mining locations in the United States, including locations in New Jersey, Pennsylvania, and Missouri. The name *zinc* was derived from the German word *zink,* but the meaning of the word *zink* is unclear.

Cadmium was discovered in 1817 by the German chemist Friedrich Stromeyer. Stromeyer found cadmium as an impurity in a sample of calamine. Recognizing that the calamine was not pure zinc oxide, he set about to discover what the impurity was and found it to be a new metal. Because the new metal was regularly found in association with zinc, the name *cadmium* was chosen, derived from *kadmeia,* the ancient Greek name for calamine.

The discovery of the only principal cadmium mineral was made in 1841 in Scotland. Consisting of cadmium sulfide (CdS), the mineral is called greenockite, after Lord Greenock (Charles Murray Cathcart [1783–1859]), its discoverer.

Mercury was one of the elements known to ancient people, especially the Chinese, Hindus, and Egyptians. It was obtained primarily

A FAMED MYTHOLOGY

Mercurius, now commonly called Mercury, was a popular ancient Roman god who was deemed responsible for much of the action and mischief that took place in antiquity. (The notion of this deity was certainly taken from the Greek god Hermes.) The first temple to Mercury was built around 495 B.C.E. in Rome. Because his role among the community of gods was to do the bidding of other, more senior gods, Mercury became known as the messenger god. A messenger personality, with enhanced abilities at communication between distant parties, would naturally be associated with trade and travel. And so Mercury became the god of trade and the protector of travelers. Words with his name at the root include mercantile, merchant, commerce, and merchandise. As the god of travelers, he also enjoyed the dubious honor of accompanying the souls of the dead to the afterlife.

Representations of Mercury usually show him with small wings attached to his sandals and hat, signifying fleetness of foot and intelligence of mind, rather than literal flight. His cleverness brought him associations with trickery in business—not

from the mineral cinnabar (HgS) and recognized to be poisonous. The Greek name for mercury was *hydrargyrum* (meaning *water silver*), from which its symbol, Hg, was obtained. Cinnabar was also ground into a fine powder and used as a pigment. Until the middle 1700s, alchemists believed that fluid nature was an essential property of mercury. In 1759, however, mercury was observed to freeze during an exceptionally cold spell in Russia.

THE CHEMISTRY OF ZINC

Zinc is essential for normal growth in both animals and plants. Without zinc, human growth and sexual development are stunted. In older males, zinc reduces enlargement of the prostate gland.

the most admirable of traits, but often useful and necessary to merchants. Another very early shrine to this deity had as its focus a fountain where, each year on Mercury's festival day (the 15th of May), merchants would pay homage and ask forgiveness for cheating their customers by sprinkling water from this fountain on their heads and wares. This may be the earliest recorded use of holy water.

The god Mercury was important enough to be honored on a particular day each week—*dies Mercurii,* or Mercury's day. The French now call it *mercredi,* the Romanians *miercuri,* the Italians *miercoledi,* and the Spanish say *miércoles.* In the English language, it is *Wednesday,* which has a relation to the Norse god, Odin, but may have no connection with Mercury.

The planet Mercury was so named for its rapid movement around the Sun. Project Mercury—the first American attempt at orbital space flight—took its name from the swift god. The name of the element mercury, initially known as *quicksilver* for its fast-flowing motion with unpredictable direction, is another of Western culture's inheritances of Roman mythology.

Metallic zinc is easily oxidized and thus is a very good reducing agent. Therefore, it is one of the most common metals used in chemistry classes to demonstrate the reaction between an active metal and hydrochloric acid. The reaction in which zinc reduces HCl to H_2 gas is shown in the following chemical equation:

$$Zn\ (s) + 2\ HCl\ (aq) \rightarrow ZnCl_2\ (aq) + H_2\ (g).$$

The hydrogen gas bubbles *(effervesces)* vigorously and can be ignited with the heat from a glowing wood splint, as shown in the following reaction:

$$2\ H_2\ (g) + O_2\ (g) \rightarrow 2\ H_2O\ (g) + heat.$$

Similar reactions occur between metallic zinc and other common acids such as sulfuric acid (H_2SO_4), dilute nitric acid (HNO_3), and acetic acid ($HC_2H_3O_2$). These acids also result in the evolution of hydrogen gas. In concentrated nitric acid, however, instead of reducing hydrogen ions, zinc reduces the nitrate ion (NO_3^-) to the ammonium ion (NH_4^+), resulting in the following reaction:

$$4\ Zn\ (s) + 10\ H^+\ (aq) + NO_3^-\ (aq) \rightarrow$$
$$4\ Zn^{2+}\ (aq) + NH_4^+\ (aq) + 3\ H_2O\ (l).$$

In this case, no gas is evolved.

The only ion zinc forms is the Zn^{2+} ion, which means that, in compounds or complex ions, zinc is present only in the "+2" oxidation state. Almost all neutral zinc compounds are white. Zinc compounds include $Zn(NO_3)_2$, $ZnCl_2$, $ZnBr_2$, ZnI_2, $ZnCO_3$, $ZnSO_4$, Zn_3P_2, $Zn_3(PO_4)_2$, ZnO, $Zn(OH)_2$, ZnS, and ZnC_2O_4. The nitrate and halides are soluble in dilute aqueous solution, while the other compounds tend to be insoluble.

$Zn(OH)_2$ is amphoteric; it dissolves in both acids and bases, as shown in the following chemical reactions:

$$Zn(OH)_2\ (s) + 2\ HCl\ (aq) \rightarrow ZnCl_2\ (aq) + 2\ H_2O\ (l);$$

$$Zn(OH)_2\ (s) + 2\ NaOH\ (aq) \rightarrow Zn(OH)_4^{2-}\ (aq) + 2\ Na^+\ (aq).$$

Other complex ions that Zn forms include $Zn(NH_3)_4^{2+}$, $Zn(CN)_4^{2-}$, and $HZnO_2^-$.

Zinc's ease of oxidation lends zinc to two important applications. The first is in common batteries such as the ones used in flashlights. Batteries operate by the occurrence of *oxidation-reduction reactions*. Zinc serves as the *anode,* or negative electrode, and is the site of oxidation. Other chemical species serve as *cathodes,* or positive electrodes, and are the sites of reduction. Electrons flow from the anode to the cathode and, in doing so, provide electrical energy that can be used for illumination or to power small electrical motors.

The second role of zinc is as a *sacrificial anode.* Zinc is a more active metal than iron is. Therefore, zinc corrodes more easily than iron does and can be used to protect iron. Galvanized nails provide an excellent example. Nails are made of steel (mostly iron), which corrodes easily when wet. By coating (galvanizing) nails with a thin layer of zinc, the zinc is oxidized (thus acting as an anode) instead of the iron, thereby extending the life of the nail far beyond what it would be without the protective zinc coating. The zinc, therefore, has been "sacrificed" to save the iron. Without the zinc coating, the iron would behave as an anode. However, because the zinc is behaving as an anode instead, the iron effectively becomes a cathode (a site of reduction). The term *cathodic protection* is used to describe the action protecting the iron by making it a cathode. (Consequently, since a neutral metal cannot in fact be reduced, nothing happens to the iron until the protective zinc coating has completely dissolved away.)

Other examples of cathodic protection occur in environments where steel is buried in soil or performs its function underwater. Gasoline stations historically stored their gasoline in underground steel tanks. Because soil gets wet, these tanks were subject to corrosion. A block of zinc attached to a tank would serve as a sacrificial anode, extending the lifetime of the tank. Today, however, many steel gasoline tanks have been replaced with thick-walled plastic tanks because of concerns over environmental degradation caused by the leakage of gasoline into the soil. Boats and ships that operate in salt water—an extremely corrosive environment—also use zinc (or sometimes magnesium) blocks as sacrificial anodes to protect the hulls. Just imagine the high cost of repairing or replacing the hull of a large oceangoing cruise ship, freighter, or oil tanker! It is much cheaper, and easier, to replace a block of zinc or magnesium than to replace the entire hull.

THE CHEMISTRY OF CADMIUM

Like zinc, cadmium exists in compounds only as a "+2" ion—Cd^{2+}. Soluble cadmium compounds include $Cd(NO_3)_2$, $CdCl_2$, and $CdBr_2$. Insoluble compounds include $CdCO_3$, CdO, $Cd(OH)_2$, and CdS. Most cadmium compounds are white, with the exceptions of $Cd(NO_3)_2$ and CdO, which are brown, and $CdBr_2$ and CdS, which are yellow.

Unlike $Zn(OH)_2$, $Cd(OH)_2$ is not amphoteric. Therefore, $Cd(OH)_2$ dissolves in acids but not in bases.

Cadmium forms a few complex ions that include $Cd(NH_3)_4^{2+}$, $Cd(CN)_4^{2-}$, and CdI_4^{2-}.

THE CHEMISTRY OF MERCURY

Liquid mercury is relatively simple to obtain from cinnabar. All that is necessary is to *roast* (heat) the ore in the presence of atmospheric oxygen. This reaction is shown in the following equation:

$$HgS \text{ (s)} + O_2 \text{ (g)} \rightarrow Hg \text{ (l)} + SO_2 \text{ (g)}.$$

Unlike its *congeners,* zinc and cadmium—which exist in compounds and complex ions only in the "+2" oxidation state—mercury exists in both the "+1" and "+2" oxidation states. The ion in the "+1" state is not entirely unique in chemistry, but it is unusual enough as to be given special attention in beginning chemistry classes. The mercury "+1" ion is called the mercurous ion and has the formula Hg_2^{2+}. What is unusual is that two metal atoms are bound together into a single ion. The ion has a charge of "+2," but each mercury atom is only in a "+1" oxidation state. Thus, in combination with a "−1" ion like Cl^-, mercurous chloride must be written as Hg_2Cl_2. The mercury ion in which mercury is in the "+2" oxidation state is called the mercuric ion and has the formula Hg^{2+}. The two ions exhibit some significantly different chemistry.

Metallic mercury itself is an extremely weak reducing agent, which means that the metal is difficult to oxidize. The mercuric ion is a fairly good oxidizing agent because it is fairly easily reduced either to the mercurous ion or to liquid mercury. The mercurous ion can undergo disproportionation. As one mercury atom in Hg_2^{2+} is oxidized to Hg^{2+}, the other atom is reduced to Hg, as shown in the following equation:

$$Hg_2^{2+} \text{ (aq)} \rightarrow Hg \text{ (l)} + Hg^{2+}.$$

The chemistry of Hg_2^{2+} is very similar to the chemistry of Ag^+. AgCl, Hg_2Cl_2, and $PbCl_2$ are all white in color and are the only three metal chlorides that are insoluble in dilute aqueous solution. ($PbCl_2$ dissolves when heated.) Mercuric chloride ($HgCl_2$), on the other hand, is soluble in aqueous solution. A common qualitative test for the presence of Hg^{2+} in solution is performed by adding stannous chloride ($SnCl_2$) to the solution. A small amount of $SnCl_2$ will reduce any Hg^{2+} present to Hg_2^{2+}, as shown in the following chemical reaction:

$$2\ Hg^{2+}\ (aq) + SnCl_2\ (aq) \rightarrow Hg_2Cl_2\ (s) + Sn^{4+}\ (aq).$$

Addition of excess $SnCl_2$ will further reduce Hg_2Cl_2 to metallic mercury, as shown in the following reaction:

$$Hg_2Cl_2\ (s) + SnCl_2\ (aq) \rightarrow 2\ Hg\ (l) + SnCl_4\ (aq).$$

In the first reaction, Hg_2Cl_2 precipitates as a white solid. In the second reaction, mercury is a silvery color. The usual result is a gray mixture that is deduced to have been caused by the presence of Hg^{2+} in the original solution.

ENVIRONMENTAL MERCURY RISKS

The accumulation of mercury waste in the environment is a worldwide challenge. Historically, a main source of contamination has been the use of mercury for gold mining. (See "Hard-rock Mining: The Environmental Costs" in chapter 5.) This practice has now fallen out of fashion, but the mercury waste remains and is washed into rivers and lakes by rainfall and snowmelt. Current sources of contamination include coal-fired power plants, crematoria, and the medical and dental industries—the latter being the largest contributors to water pollution by mercury. In dentistry, the material for filling cavities in teeth has for decades been made of an amalgam containing about 50 percent mercury. Wastewater containing dental amalgam material or medical waste such as mercury-containing thermometers, blood pressure and gastrointestinal devices, laboratory barometers, and mercury fixatives is mostly unmonitored, but widespread. *Sphygmomanometers,* instruments that measure blood pressure, constitute the single largest use of mercury in the medical industry.

Mercury is a heavy metal, tending to settle to the bottoms of lakes and wetlands, rather than being washed far out to sea. It accumulates over time and is absorbed by aquatic life, becoming more concentrated higher in the food chain. Pregnant women have long been warned to restrict their fish consumption for this reason. Inhalation of mercury vapors, which are colorless and odorless, is even more hazardous than ingestion. Dental workers are therefore at risk, as are those who work in industries that use mercury in their manufacturing processes. The production of felt hats, which were quite popular in the early to mid-1900s, relied on the use of mercury in preparation. Hatmakers frequently displayed odd personality characteristics, leading to the phrase "mad as a hatter." The problem was eventually recognized, and the use of mercury in hatmaking was banned in 1941. Likewise, mercury thermometers (which easily break) in schools and elsewhere have now been widely replaced by alcohol-containing thermometers.

The symptoms of mercury poisoning are multiple and frequently improperly diagnosed. They include seemingly mild effects such as skin rashes, tingling sensations in hands and feet, digestive problems, and insomnia, as well as more severe problems like paralysis, speech impairment, kidney damage, insanity, and possible death. Even small exposures may have very damaging health effects. The World Health Organization has stated that there may be "no threshold below which some adverse effects do not occur."

In 2003, the United Nations Environment Programme (UNEP) instigated a global mercury partnership with the ultimate goal of eliminating all *anthropogenic* mercury releases. The partnership held its first meeting in April 2009. Suggestions for mercury reduction include improvement of waste management, investigations into ways of cleaning up coal emissions, and finding alternatives to mercury in medical equipment. For dental use, composite resins—now available as reliable replacement material for mercury-silver amalgams—are advised. As of September 2009, there were 46 partners in the UNEP global collaboration, including the World Medical Association, the National Resources Defense Council, the World Dental Federation, the Artisanal Gold Council, and the Association for Responsible Mining.

TECHNOLOGY AND CURRENT USES OF ZINC, CADMIUM, AND MERCURY

As a relatively abundant element, zinc has a number of uses that include the following. Zinc oxide is the principal ingredient in calamine lotion,

Before the advent of advanced sun-block creams, white zinc oxide cream was commonly used to protect the nose against sunburn. *(Carl Mydans/Time Life Pictures/Getty Images)*

which is used to help heal wounds. Zinc is used in batteries, especially AAA, AA, C, and D batteries. Zinc is present in the alloys used to make coins. Zinc is used in cathodic protection of steel and is alloyed with copper to make brass. Because zinc compounds tend to be white in color, they can be used as pigments. Beginning around 1850, inexpensive metallic zinc was widely used in sculpture in the United States, but it fell out of favor in the mid-20th century.

Uses of cadmium are limited because of cadmium's toxicity and carcinogenicity. What applications of cadmium do exist include the following. Cadmium is probably best known for its use with nickel in rechargeable *Ni-Cad batteries*. Cadmium is alloyed with silver to form a low-melting point solder. Steel can be coated with a thin layer of cadmium to minimize corrosion. Cadmium is one choice of metal used in control rods in nuclear reactors.

Several cadmium compounds have applications. Cadmium bromide is used in film photography. Because of its yellow color, cadmium chloride is used in oil paints. Cadmium compounds are used in television and computer screens. Cadmium sulfate finds a number of uses in fluorescent screens, paints, ceramic glazes, photovoltaic cells, and fire-works. Copper can be hardened by alloying it with cadmium.

Historically, the use of cadmium as a coating on iron and steel to prevent corrosion was the major application of cadmium. Today, to a large extent, zinc has replaced cadmium because of environmental concerns about the exposure of workers to cadmium during the electroplating process and the problems associated with subsequent waste disposal. The use of cadmium in Ni-Cad batteries, however, continues to be important, as does its use in detectors for medical imaging.

Mercury is probably best known for its use in thermometers. However, it is also commonly used to make amalgams (solid solutions) with other metals. Several metals, including gold, silver, platinum, and copper, form amalgams. The most familiar amalgam is one containing mercury and silver, which is still used by some dentists to fill cavities.

Because mercury is a good conductor of electricity, it is used as a liquid contact in certain electrical switches. Streetlights sometimes contain mercury vapor. Mercury is used in barometers. Mercury is also used in the electrolysis of *brine* (seawater) to produce chlorine gas and sodium hydroxide. Mercury has also been added to paints to prevent mildew. Some mercury is used in pulp and paper manufacturing processes and in the manufacture of compact fluorescent light bulbs.

Several mercury compounds have applications. Mercuric cyanate $(Hg[CNO]_2)$ is used in explosives. Mercuric chloride $(HgCl_2)$ is used as a disinfectant. A mercury ointment $(HgNH_2Cl)$ is used to treat certain skin diseases. Compounds of mercury are used in herbicides and pesticides. To at least some degree, the use of mercury is being phased out because of its toxicity. Other substances are being substituted to minimize exposure to humans.

7

Conclusions and Future Directions

The periodic table of the elements is a marvelous tool, one that scientists have only begun to investigate. The key to its utility is its organization and the patterns it weaves. It can guide the eye and the mind to understand how far science has come and where human knowledge can lead.

SPECULATIONS ON FUTURE DEVELOPMENTS

It is important for scientists to think ahead, to attempt to guess what areas are ripe for investigation and which may be bound for oblivion. If one considers the recent remarkable leaps in information science, medicine, and particle physics that have materialized in the past century, it is clear that predictions are bound to be less impressive than eventualities, but there are some obvious starting points. Some of those especially related to transition metals are suggested here.

NEW PHYSICS

The near future of the transition metals will most likely center on the strengthening and corrosion resistance of materials, enhancement of computer technology, and considerations regarding the environment and the promotion of cleaner energy. In particular, some environmental hazards need to be addressed in a timely manner, especially as regards human health.

Problems with mining waste and toxic releases abound while legislation moves at a pace that does not keep up with the needs of human health or the environment. Recent cyanide spills from gold mines have devastated entire cities and river ecosystems. Ways to minimize acid mine drainage and clean up past mercury waste will be an ongoing challenge. While mercury has commonly been acknowledged as toxic, legislation is not forthcoming regarding cleanup or regulation. The United Nations Environment Programme's global mercury partnership is a good start, but six years passed between the agreement to work on the problem and the first meeting to address it. The health risks of tungsten and cadmium also need to be investigated more rigorously.

Regarding new energy sources, the investigation of low-energy nuclear reactions will continue to be of interest, as will the superconducting nature of triniobium tin and other new alloys.

Titanium's mechanical properties make it an excellent structural material, particularly for the aerospace industry. Not only aircraft manufacturers, but also makers of sports equipment, automobiles, pressure vessels, and many others would like to incorporate more titanium in their designs. Unfortunately, the element in its pure form is difficult and expensive to extract, process, and fabricate. For an identical end product, an order of magnitude increase in cost is typical when compared with stainless steel. Researchers at the U.S. Department of Defense, among others, are investigating ways to reduce the production expense. Other experiments at strengthening material involve reducing the size of grains of one type of element embedded in the metallic lattice of another material. New alloys may be created with greatly enhanced properties.

In computer science, minimizing the size of the chips is always at the forefront of research. The switch to hafnium-based semiconductors may facilitate size reduction, but there are problems with electrode compatibility that may make the transition costly in the short run, but research and development will continue.

Regarding the science behind understanding the formation of transition metals in stars, much remains to be discovered. In particular, continuing research on the excesses of heavy metals in chemically peculiar stars will advance knowledge about the thermodynamics and mass-dependence of nucleosynthesis of the elements.

NEW CHEMISTRY

Transition metals are the workhorses of industry. As long as world deposits are economically recoverable, transition metals will continue to be in high demand. The economics of mining have two aspects. The first is the cost itself of exploration, discovery of deposits, extraction, and transportation to refineries and other processing facilities. The second is the metals' market values. Metals are commodities, and their prices rise and fall according to the basic principles of economic supply and demand. It has frequently been the case, for example, that a gold mine has been shut down by its owners because the recovery of the remaining gold became too expensive to justify continuing mining operations. Subsequently, mining operations have been resumed either when more economical means of recovery have been developed, or when the price of gold has risen, or possibly when both events have occurred. All current indicators suggest that the demand for metals is strong, that prices reflect supply and demand, and that many modern industries will continue to require metals for their continuing operations. Two questions suggest themselves, however. The first question is: What new transition metal chemistry can be anticipated in coming decades? The second question is: Will new elements be discovered that will be classified as transition metals?

The answer to the first question lies in the continued development and improvement of existing industrial processes. The importance of steel—and the accompanying demand for iron and metals like chromium, manganese, molybdenum, and tungsten—is fairly well known. Also well known are the use of copper and nickel in coins minted for general circulation, and the uses of silver, gold, and platinum in commemorative coinage, bullion, and jewelry. Many people may be familiar with the use of platinum group metals in the catalytic converters of motor vehicles.

An extremely important area of transition metal chemistry, however, with which most people probably are *not* familiar is the metals' uses in the form of *coordination compounds*. In coordination compounds, other groups of atoms, called ligands, typically bond to the metal ions in multiple positions. Some of these combinations of metal ions and ligands can be quite complex, so another term that is used to describe them is to call them *metal complexes* or *sandwich compounds*. The composition of these coordination compounds fundamentally alters the properties of the metal ions, allowing them to undergo much greater varieties of chemical reactions. Coordination compounds have important applications in catalysis. It may be stated with confidence that many of the new developments to be expected in transition metal chemistry in the coming decades will be in the area of finding more efficient and more economical chemical catalysts.

In regard to the question about the discovery of new transition metal elements, the answer is that 10 new elements have already been discovered. Actinium (element 89) and the first 10 translawrencium elements (elements with atomic numbers greater than 103)—elements 104 through 112—have already been added to the periodic table. The first row of transition metals—scandium through zinc—is characterized by electrons filling the atoms' 3d subshell. The second row of transition metals—yttrium through cadmium—is characterized by electrons filling the 4d subshell. The third row—consisting of lanthanum through mercury—is characterized by electrons filling the 5d subshell. Consequently, the fourth row—consisting of element 104 (rutherfordium) through element 112 (copernicium)—is characterized by electrons filling the 6d subshell.

Because elements 104–112 are artificially made elements, the syntheses of which follow the same techniques as the other *transuranium elements*, elements 104–112 are referred to as *transactinide* elements and are not included in this book. The question about the discovery of new transition metals thus becomes: Will elements with electrons in the 7d subshell ever be discovered? If they ever are discovered, it would be required that their atomic numbers lie in the range of 130–140. It simply is unknown at this time whether or not it will ever be possible to produce elements that are that heavy. An entire additional row of rare earth elements would have to be discovered first. Extremely high energies would be required to produce such heavy elements, requiring technologies that presently do not exist. Any prospect of ever doing so undoubtedly lies many years into the future. But that very uncertainty is what makes the quest for scientific knowledge so exciting. What seems highly improbable or even impossible today could become common-place tomorrow.

SI Units and Conversions

UNIT	QUANTITY	SYMBOL	CONVERSION
Base units			
meter	length	m	1 m = 3.2808 feet
kilogram	mass	kg	1 kg = 2.205 pounds
second	time	s	
ampere	electric current	A	
kelvin	thermodynamic temperature	K	1 K = 1°C = 1.8°F
candela	luminous intensity		
mole	amount of substance	mol	
Supplementary units			
radian	plane angle	rad	pi / 2 rad = 90°
steradian	solid angle	sr	
Derived units			
coulomb	quantity of electricity	C	
cubic meter	volume	m^3	$1 \ m^3 = 1.308 \ yards^3$
farad	capacitance	F	
henry	inductance	H	
hertz	frequency	Hz	
joule	energy	J	1 J = 0.2389 calories
kilogram per cubic meter	density	$kg \ m^{-3}$	$1 \ kg \ m^{-3} = 0.0624 \ lb. \ ft^{-3}$
lumen	luminous flux	lm	
lux	illuminance	lx	
meter per second	speed	$m \ s^{-1}$	$1 \ m \ s^{-1} = 3.281 \ ft \ s^{-1}$

UNIT	QUANTITY	SYMBOL	CONVERSION
meter per second squared	acceleration	$m\ s^{-2}$	
mole per cubic meter	concentration	$mol\ m^{-3}$	
newton	force	N	1 N = 7.218 lb. force
ohm	electric resistance	Ω	
pascal	pressure	Pa	$1\ Pa = \dfrac{0.145\ lb}{in^{-2}}$
radian per second	angular velocity	$rad\ s^{-1}$	
radian per second squared	angular acceleration	$rad\ s^{-2}$	
square meter	area	m^2	$1\ m^2 = 1.196\ yards^2$
tesla	magnetic flux density	T	
volt	electromotive force	V	
watt	power	W	$1W = 3.412\ Btu\ h^{-1}$
weber	magnetic flux	Wb	

PREFIXES USED WITH SI UNITS		
PREFIX	SYMBOL	VALUE
atto	a	$\times 10^{-18}$
femto	f	$\times 10^{-15}$
pico	p	$\times 10^{-12}$
nano	n	$\times 10^{-9}$
micro	μ	$\times 10^{-6}$
milli	m	$\times 10^{-3}$
centi	c	$\times 10^{-2}$
deci	d	$\times 10^{-1}$
deca	da	$\times 10$
hecto	h	$\times 10^{2}$
kilo	k	$\times 10^{3}$
mega	M	$\times 10^{6}$
giga	G	$\times 10^{9}$
tera	T	$\times 10^{12}$

List of Acronyms

AGB	asymptotic giant branch
BCC	body-centered cubic
CNO	carbon-nitrogen-oxygen
CP	chemically peculiar
EPA	Environmental Protection Agency
FCC	face-centered cubic
FDA	Food and Drug Association
HCP	hexagonal, close-packed
ISM	interstellar medium
IUPAC	International Union of Pure and Applied Chemistry
LHC	Large Hadron Collider
MeV	million electron volts
PG&E	Pacific Gas & Electric
ppb	parts per billion
SneI	Type-I supernovae
SneII	Type-II supernovae
UNEP	United Nations Environment Programme
USGS	United States Geological Survey
YAG	yttrium-aluminum-garnet

Periodic Table of the Elements

Periodic Table of the Elements

Legend: Atomic number — Symbol — Atomic weight

3 / Li / 6.941

Halogens
Metals
Nonmetals
Metalloids
Unknown

Numbers in parentheses are atomic mass numbers of most stable isotopes.

1 IA	2 IIA	3 IIIB	4 IVB	5 VB	6 VIB	7 VIIB	8 VIIIB	9 VIIIB	10 VIIIB	11 IB	12 IIB	13 IIIA	14 IVA	15 VA	16 VIA	17 VIIA	18 VIIIA
1 H 1.00794																	2 He 4.0026
3 Li 6.941	4 Be 9.0122											5 B 10.81	6 C 12.011	7 N 14.0067	8 O 15.9994	9 F 18.9984	10 Ne 20.1798
11 Na 22.9898	12 Mg 24.3051											13 Al 26.9815	14 Si 28.0855	15 P 30.9738	16 S 32.067	17 Cl 35.4528	18 Ar 39.948
19 K 39.0938	20 Ca 40.078	21 Sc 44.9559	22 Ti 47.867	23 V 50.9415	24 Cr 51.9962	25 Mn 54.938	26 Fe 55.845	27 Co 58.9332	28 Ni 58.6934	29 Cu 63.546	30 Zn 65.409	31 Ga 69.723	32 Ge 72.61	33 As 74.9216	34 Se 78.96	35 Br 79.904	36 Kr 83.798
37 Rb 85.4678	38 Sr 87.62	39 Y 88.906	40 Zr 91.224	41 Nb 92.9064	42 Mo 95.94	43 Tc (98)	44 Ru 101.07	45 Rh 102.9055	46 Pd 106.42	47 Ag 107.8682	48 Cd 112.412	49 In 114.818	50 Sn 118.711	51 Sb 121.760	52 Te 127.60	53 I 126.9045	54 Xe 131.29
55 Cs 132.9054	56 Ba 137.328	57-70 ☆	71 Lu 174.967	72 Hf 178.49	73 Ta 180.948	74 W 183.84	75 Re 186.207	76 Os 190.23	77 Ir 192.217	78 Pt 195.08	79 Au 196.9655	80 Hg 200.59	81 Tl 204.3833	82 Pb 207.2	83 Bi 208.9804	84 Po (209)	85 At (210)	86 Rn (222)
87 Fr (223)	88 Ra (226)	89-102 ★	103 Lr (260)	104 Rf (261)	105 Db (262)	106 Sg (266)	107 Bh (262)	108 Hs (263)	109 Mt (268)	110 Ds (271)	111 Rg (272)	112 Cn (277)	113 Uut (284)	114 Uuq (285)	115 Uup (288)	116 Uuh (292)	118 Uuo (294)

☆ Lanthanides

57 La 138.9055	58 Ce 140.115	59 Pr 140.908	60 Nd 144.24	61 Pm (145)	62 Sm 150.36	63 Eu 151.966	64 Gd 157.25	65 Tb 158.9253	66 Dy 162.500	67 Ho 164.9303	68 Er 167.26	69 Tm 168.9342	70 Yb 173.04

★ Actinides

89 Ac (227)	90 Th 232.0381	91 Pa 231.036	92 U 238.0289	93 Np (237)	94 Pu (244)	95 Am 243	96 Cm (247)	97 Bk (247)	98 Cf (251)	99 Es (252)	100 Fm (257)	101 Md (258)	102 No (259)

© Infobase Publishing

Table of the Elements Categories

Element Categories

Nonmetals
1	H	Hydrogen
6	C	Carbon
7	N	Nitrogen
8	O	Oxygen
15	P	Phosphorus
16	S	Sulfur
34	SE	Selenium

Halogens
9	F	Fluorine
17	Cl	Chlorine
35	Br	Bromine
53	I	Iodine
85	At	Astatine

Noble Gases
2	He	Helium
10	Ne	Neon
18	AT	Argon
36	Kr	Krypton
54	Xe	Xenon
86	Ra	Radon

Metalloids
5	B	Boron
14	Si	Silicon
32	Ge	Germanium
33	As	Arsenic
51	Sb	Antimony
52	Te	Tellurium
84	Po	Polonium

Alkali Metals
3	Li	Lithium
11	Na	Sodium
19	K	Potassium
37	Rb	Rubidium
55	Cs	Cesium
87	Fr	Francium

Alkaline Earth Metals
4	Be	Beryllium
12	Mg	Magnesium
20	Ca	Calcium
38	Sr	Strontium
56	Ba	Barium
88	Ra	Radium

Post-Transition Metals
13	Al	Aluminum
31	Ga	Gallium
49	In	Indium
50	Sn	Tin
81	Tl	Thallium
82	Pb	Lead
83	Bi	Bismuth

Transactinides
104	Rf	Rutherfordium
105	Db	Dubnium
106	Sg	Seaborgium
107	Bh	Bohrium
108	Hs	Hassium
109	Mt	Meitnerium
110	Ds	Darmstadtium
111	Rg	Roentgenium
112	Uub	Ununbium
113	Uut	Ununtrium
114	Uuq	Ununquadium
115	Uup	Ununpentium
116	Uuh	Ununhexium
118	Uuo	Ununoctium

Transition Metals
21	Sc	Scandium	39	Y	Yttrium	72	Hf	Hafnium
22	Ti	Titanium	40	Zr	Zirconium	73	Ta	Tantalum
23	V	Vanadium	41	Nb	Niobium	74	W	Tungsten
24	Cr	Chromium	42	Mo	Molybdenum	75	Re	Rhenium
25	Mn	Manganese	43	Tc	Technetium	76	Os	Osmium
26	Fe	Iron	44	Ru	Ruthenium	77	Ir	Iridium
27	Co	Cobalt	45	Rh	Rhodium	78	Pt	Platinum
28	Ni	Nickel	46	Pd	Palladium	79	Au	Gold
29	Cu	Copper	47	Ag	Silver	80	Hg	Mercury
30	Zn	Zinc	48	Cd	Cadmium			

Note: The organization of the periodic table of the elements, while useful to chemists and physicists, may be confusing to non-scientists in that some groupings of similar elements appear as vertical columns (halogens, for example); some as horizontal rows (lanthanides, for example); and some as a combination of both (nonmetals). The table of element categories is intended as a quick reference sheet to easily determine which elements belong to which groups. (Element 117 does not appear in this list because it is undiscovered as of the publishing of this book.)

Lanthanides
57	La	Lanthanum	62	Sm	Samarium	67	Ho	Holmium
58	Ce	Cerium	63	Eu	Europium	68	Er	Erbium
59	Pr	Praseodymium	64	Gd	Gadolinium	69	Tm	Thulium
60	Nd	Neodymium	65	Tb	Terbium	70	Yb	Ytterbium
61	Pm	Promethium	66	Dy	Dysprosium	71	Lu	Lutedum

Actinides
89	Ac	Actinium	94	Pu	Plutonium	99	Es	Einsteinium
90	Th	Thorium	95	Am	Americium	100	Fm	Fermium
91	Pa	Protactinium	96	Cm	Curium	101	Md	Mendelevium
92	U	Uranium	97	Bk	Berkelium	102	No	Nobelium
93	Np	Neptunium	98	Cf	Californium	103	Lr	Lawrencium

Chronology

Ancient times	Civilizations are using copper, silver, and gold in coins.
ca. 3000–2500 B.C.E.	Civilizations begin reducing iron oxide to elemental iron.
ca. 1250 B.C.E.	Cobalt is used in glass coloration in Mesopotamia and Egypt.
ca. 1000 B.C.E.	Iron smiths in the Middle East begin producing steel.
495 B.C.E.	First known temple to the Roman god, Mercury, is dedicated.
ca. 1200s C.E.	In China and India, metallic zinc is being smelted from its ores.
1694	Swedish chemist and mineralogist Georg Brandt is born on July 21 in Skinnskatteberg, Sweden.
1709	German chemist Andreas Sigismund Marggraf is born on March 3 in Berlin.
1716	Don Antonio de Ulloa is born on January 12 in Seville, Spain.
1722	Swedish chemist Axel Fredrik Cronstedt is born on December 23 in Södermanland, Sweden.
1734	Antoine Grimoald Monnet is born in Champeix, Auvergne, France.
1735	Swedish chemist Torbern Bergman is born on March 20 in Katrineberg, Sweden.
	Georg Brandt announces discovery of cobalt.
1740	French chemist Balthasar-Georges Sage is born in Paris.
1742	Swedish chemist Carl Wilhelm Scheele is born on December 9 in Stralsund, Sweden.
1743	German chemist Martin Heinrich Klaproth is born on December 1 in Wernigerode, Germany.
1744	Don Antonio de Ulloa and Jorge Juan complete the research that results in the discovery of platinum.

1777 Danish chemist Hans Christian Ørsted is born on August 14 in Rudkøbing, Denmark.

1745 Swedish chemist Johan Gottlieb Gahn is born on August 19 in Voxna, near Söderhamn, Sweden.

1746 Swedish chemist Peter Jacob Hjelm is born on October 2 in Sunnerbo Härad, Småland, Sweden.

Andreas Marggraf isolates zinc.

1751 Axel Fredrik Cronstedt announces discovery of nickel.

1755 French chemist Antoine-François de Fourcroy is born on June 15 in Paris.

1760 Finnish chemist Johan Gadolin is born on June 5 in Abo, Finland.

1761 English chemist Smithson Tennant is born on November 30 near Richmond, Wensleydale, England.

Reverend William Gregor, an amateur chemist and mineralogist, is born on December 25 in Trewarthenick, Cornwall, England.

1763 French chemist Nicolas-Louis (or Louis Nicolas) Vauquelin is born on May 16 in Saint-Andrée d'Hebertut, near Pont l'Évêque, Normandy.

1764 Spanish-Mexican mineralogist Andrés Manuel del Río is born on November 10 in Madrid, Spain.

1765 English chemist Charles Hatchett is born on January 2 in London.

Axel Fredrik Cronstedt dies on August 19 in Stockholm, Sweden.

1766 English chemist and physicist William Hyde Wollaston is born on August 6 in East Dereham, England.

1767 Swedish chemist and mineralogist Anders Gustaf Ekeberg is born on January 16 in Stockholm, Sweden.

1768 Georg Brandt dies on April 29 in Stockholm, Sweden.

Polish chemist Andrei Sniadecki is born on November 30 in Zinn, Lithuania.

1778   Carl Wilhelm Scheele isolates molybdic acid.

1779   Swedish chemist Jöns Jacob Berzelius is born on August 20 in Vaversunda, Sweden.

1782   Andreas Sigismund Marggraf dies on August 7 in Berlin, Germany.

1783   Lord Greenock (Charles Murray Cathcart) is born on December 21 in Walton, Essex, England.

1784   Torbern Bergman dies on July 8 in Medevi, Sweden.

1786   Carl Wilhelm Scheele dies on May 21, most likely in Köping, Sweden.

1787   Swedish physician and chemist Nils Gabriel Sefström is born on June 2 in Ilsbo, Hälsingland, Sweden.

1794   Johan Gadolin announces discovery of yttrium.

1795   Don Antonio de Ulloa dies on July 3 in Isla de Leon in Cadiz, a province of Spain.

       German analytical chemist Heinrich Rose is born on August 6 in Berlin.

1796   Karlovich Klaus is born on January 23 in Dorpat Russia (now Estonia).

1800   German chemist Friedrich Wöhler is born on July 31 in Eschersheim, Frankfurt am Main, Germany.

1804   William Hyde Wollaston succeeds in isolating rhodium from platinum ore.

1809   Antoine-Francois de Fourcroy dies on December 16 in Paris, France.

1813   Anders Gustaf Ekeberg dies on February 11 in Uppsala, Sweden.

       German metallurgist C. J. A. Theodor Scheerer is born on August 28 in Berlin.

       Peter Jacob Hjelm dies on October 7 in Stockholm, Sweden.

1815   Smithson Tennant dies on February 22 in Boulogne, France.

1817   German chemist Friedrich Stromeyer announces discovery of cadmium.

Martin Klaproth dies on January 1 in Berlin, Germany.

Antoine Monnet dies on May 23 in Paris, France.

William Gregor dies on June 11 in Creed, Cornwall before completing the work that was leading to the discovery of titanium.

1818   Johan Gottlieb Gahn dies on December 8 in Stockholm, Sweden.

1820   Hans Christian Øersted discovers that strong magnetic fields can be produced by winding conducting wire around a cylinder and then passing a current through the wire.

1824   Berzelius succeeds in obtaining an impure sample of zirconium.

Balthasar-Georges Sage dies in France.

1828   William Wollaston dies on December 22 in London, England.

1829   Nicolas-Louis Vauquelin dies on November 14 near Pont l'Eveque, Normandy.

1832   English chemist Sir William Crookes is born on June 17 in London, England.

1833   English chemist Henry Roscoe is born on January 7 in London, England.

1834   Russian chemist Dmitri Mendeleev is born on February 8 in Siberia.

1838   Polish chemist Andrei Sniadecki dies on May 12 in Vilnius, Lithuania.

1840   Swedish analytical chemist Lars Fredrik Nilson is born on May 14 in Skönberga parish in Österötland, Sweden.

1841   Lord Greenock (Charles Murray Cathcart) discovers cadmium sulfide.

1845   Nils Gabriel Sefström dies on November 30 in Stockholm, Sweden.

1847 Charles Hatchett dies on March 10 in London, England.

1848 Jöns Jacob Berzelius dies on August 7 in Stockholm, Sweden.

1849 California gold rush begins.

Andrés Manuel del Río dies on March 23 in Mexico City.

1851 Danish physicist Hans Christian Ørsted dies on March 9 in Copenhagen, Denmark.

1852 Johan Gadolin dies on August 15 in Wirmo, Finland.

1857 Lord Greenock (Charles Murray Cathcart) dies on July 16 in St. Leonard's-on-Sea, England.

1864 Heinrich Rose dies on January 27 in Germany.

Karlovich Klaus dies on March 24 in Dorpat, Russia.

1867 Henry Roscoe isolates pure metallic vanadium.

1869 Dmitri Mendeleev publishes his periodic table of the elements.

1873 Scientist Otto Berg is born in Germany.

1875 Theodor Scheerer dies on July 19 in Dresden, Germany.

1879 German physicist Max von Laue is born on October 9 in Pfaffendorf, near Koblenz, Germany.

1882 Friedrich Wöhler dies on September 23 in Göttingen, Germany.

1885 Danish physicist Niels Bohr is born on October 7 in Copenhagen, Denmark.

1886 Italian chemist Carlo Perrier is born in July in Turin, Italy.

1887 English physicist Henry Moseley is born on November 23 in Weymouth, Dorset.

1889 Hungarian chemist George Charles de Hevesy is born on August 1 in Budapest, Hungary.

Dutch physicist Dirk Coster is born on October 5 in Amsterdam, the Netherlands.

1893 German chemist Walter Noddack is born on August 17 in Berlin, Germany.

1896 German chemist Ida Tacke is born on February 25 in Lackhausen, Germany.

1899 Lars Fredrik Nilson dies on May 14 in Stockholm, Sweden.

1901 American physicist Ernest Orlando Lawrence is born on August 8 in Canton, South Dakota.

1905 Italian-American physicist Emilio Segrè is born on February 1, in Rome, Italy.

1907 Dmitri Mendeleev dies on February 2 in St. Petersburg, Russia.

1908 Sir William Crookes discovers that scandium occurs widely on Earth.

1912 Max von Laue discovers a method for measuring frequencies of X-rays.

1913 Moseley demonstrates that the positions of elements in the periodic table should be in order of atomic number, not atomic weight.

1914 Max von Laue receives Nobel Prize in physics for the discovery of diffraction of X-rays in crystals.

1915 Henry Moseley is killed in battle on August 10 in Gallipoli, Turkey.

1919 Henry Roscoe dies on December 18 in London, England.

Sir William Crookes dies on April 4 in London, England.

1923 George de Hevesy and Dirk Coster use X-ray analysis to identify hafnium.

1925 Walter Noddack, Ida Tacke, and Otto Berg report the discovery of rhenium.

1939 De Hevesy is elected a Friend of the Royal Society.

Ernest Orlando Lawrence receives the Nobel Prize in physics for his invention of the cyclotron.

Otto Berg dies in Germany.

1943   De Hevesy receives the Nobel Prize in chemistry for his work on the use of isotopes as tracers in the study of chemical processes.

1948   Carlo Perrier dies in Genoa, Italy.

1950   Dirk Coster dies on February 12 in Groningen, the Netherlands.

1951   The International Union of Pure and Applied Chemistry officially declares the name of element 41 to be *niobium*.

1958   Ernest Orlando Lawrence dies on August 27 in Palo Alto, California.

1959   De Hevesy is awarded the Atoms for Peace Award.

1960   Max von Laue dies on April 24 in Berlin, Germany.

       Walter Noddack dies on December 7 in Berlin.

1962   Niels Bohr dies on December 18 in Copenhagen, Germany.

1966   George de Hevesy dies.

ca. 1970–80   East German secret police employ radioactive scandium 46 for surveillance of human subjects.

1978   Ida Noddack (née Ida Tacke) dies on October 29 in Bad Neuenahr, Germany.

1989   Emilio Segrè dies on April 22 in Lafayette, California.

       Stanley Pons and Martin Fleischmann announce on March 23 their discovery of a way to produce fusion energy on a tabletop at room temperature.

1994   Ore of rhenium sulfide is discovered in Russia

1996   Environmental activist, Erin Brockovich, wins case against PG&E for releasing water containing hexavalent chromium into unlined ponds.

1998   The discovery of element 114 (as yet unnamed) is announced by workers at the Nuclear Institute at Dubna, Russia.

2000   The discovery of element 116 (as yet unnamed) is announced by workers at the Nuclear Institute at Dubna, Russia.

2003   The discoveries of elements 113 and 115 (both as yet unnamed) are announced by workers at the Nuclear Institute at Dubna, Russia.

2005   Archaeologists report the discovery in Qantir, Egypt, of an ancient glass-manufacturing plant that produced cobalt blue ingots.

2008   French scientists report the first successful synthesis of a nickel 56 nucleus.

The Large Hadron Collider begins proton beam circulation on September 10.

California governor Arnold Schwarzenegger signs legislation banning lead ammunition.

2010   On March 10 the LHC successfully collides two proton beams at record high energies (7 trillion electron volts).

The discovery of element 117 (as yet unnamed) is announced by researchers at the Joint Institute for Nuclear Research in Dubna, Russia.

Glossary

accretion   referring to a two-star system, the tendency of the more massive star to attract matter from its companion star.

acid   a type of compound that contains hydrogen and dissociates in water to produce hydrogen ions.

acid mine drainage   acidic water that drains from mining facilities into the surrounding soil and aquifers.

acoustic   relating to sound and sound waves.

actinides   the elements ranging from thorium (atomic number 90) to lawrencium (number 103); all have two outer electrons in the 7s subshell plus an increasing number of electrons in the 5f subshell.

alkali metal   the elements in column IA of the periodic table (exclusive of hydrogen); all are characterized by a single valence electron in an *s* subshell.

alkaline earth metal   the elements in column IIA of the periodic table; all are characterized by two valence electrons that fill an *s* subshell.

alpha decay   a mode of radioactive decay in which an alpha particle is emitted. The daughter isotope has an atomic number two units less than the atomic number of the parent isotope, and a mass number that is four units less.

alpha particle   a nucleus of helium 4.

alpha process   a process by which helium nuclei bond with other nuclei in the cores of stars to make heavier elements with nuclear numbers that are a multiple of 4.

amalgam   a mixture of mercury with another metal.

ambient   surrounding.

amino acid   an organic compound that contains an $-NH_2$ group and a $-COOH$ group.

amphoteric   referring to a substance that can react with both acids and bases.

analytical chemistry   the branch of chemistry that determines the chemical constitution of a chemical sample. (See *qualitative analysis* and *quantitative analysis.*)

anemia   a medical disorder characterized by a low level of healthy red blood cells.

anion   an atom with one or more extra electrons, giving it a net negative charge.

anthropogenic   generated by humans.

antiseptic   able to clean or disinfect; literally "against sepsis."

anode   the site of oxidation in an electrochemical cell. In an electrolytic cell, the anode is positively charged; in a galvanic cell, the anode is negatively charged.

apothecary   a druggist or pharmacist.

aqua regia   latin for "royal water"; a mixture of concentrated hydrochloric and nitric acid in a 3:1 ratio.

aqueous   describing a solution in water.

aquifer   an underground body of water surrounded by rock or soil.

aromatic hydrocarbon   a compound that contains groups of six carbon atoms arranged in hexagons.

asymptotic giant branch (AGB)   an area of the *Hertzsprung-Russell diagram* above the main sequence line, where some high-mass stars (AGB stars) are mapped for luminosity and temperature.

atom   the smallest part of an element that retains the element's chemical properties; atoms consist of protons, neutrons, and electrons.

atomic mass   the mass of a given isotope of an element—the combined masses of all its protons, neutrons, and electrons.

atomic number   the number of protons in an atom of an element; the atomic number establishes the identity of an element.

atomic weight   the mean weight of the atomic masses of all the atoms of an element found in a given sample, weighted by isotopic abundance.

autooxidation   (See *disproportionation.*)

band   an energy level that an electron can occupy in a crystalline solid.

base   a substance that reacts with an acid to give water and a salt; a substance that, when dissolved in water, produces hydroxide ions.

beta decay   a mode of radioactive decay in which a beta particle—an ordinary electron—is emitted; the daughter isotope has an atomic number one unit greater than the atomic number of the parent isotope, but the same mass number.

binary stars   two stars that are close enough to each other that they both revolve around their mutual center of mass.

blast furnace   a furnace used for smelting iron ores to make pig iron.

blue compact galaxies   relatively dust-free galaxies containing a high proportion of hot, young stars that give off blue light.

brine   water containing large amounts of salt, usually sodium chloride; salt water.

bullion   a bar or ingot of a precious metal such as gold, silver, or platinum.

carbon-nitrogen-oxygen cycle   one of the mechanisms of nuclear fusion by which hydrogen is converted into helium.

carcinogenic   describing a substance that produces cancer.

carrion   the remains of a dead animal.

catalyst   a chemical substance that speeds up a chemical reaction without itself being consumed by the reaction.

catalytic converter   a device in motor vehicles that uses a catalyst—usually platinum or another platinum group metal—to reduce emissions of pollutant gases.

cathode   the site of reduction in an electrochemical cell. In an electrolytic cell, the cathode is negatively charged; in a galvanic cell, the cathode is positively charged.

cathodic protection   the technique of protecting a metal from corrosion by making it serve as a cathode in an electrochemical cell; to corrode, the metal would have to act as an anode.

cation an atom that has lost one or more electrons to acquire a net positive charge.

Chandrasekhar mass the upper limit to the mass of a white dwarf star.

chemical bond the force of attraction holding atoms together in molecules or crystals.

chemical change a change in which one or more chemical elements or compounds form new compounds; in a chemical change, the names of the compounds change.

chemically peculiar star any star that shows anomalies in the expected abundance of various elements, usually having a high proportion of heavy elements as compared to solar abundance.

cloud seeding a method of increasing rain or other precipitation by adding substances to the atmosphere that cause water droplets to form.

CNO (See *carbon-nitrogen-oxygen cycle*.)

cobalt blue a deep blue pigment made from cobalt compounds.

cofactor a nonprotein chemical substance that is essential for the normal function of an enzyme.

coke coal that has had impurities removed.

cold-worked a method to shape steel that is at a temperature where it is no longer flexible, by rolling or hammering.

columbite an important mineral that contains niobium.

complex ion any ion that contains more than one atom.

complex salt an ionic compound that contains more than one positive ion.

composite a material that combines the strengths and abilities of two different materials; composites have been in use for thousands of years.

compound a pure chemical substance consisting of two or more elements in fixed, or definite, proportions.

congener an element that belongs to the same group as another element; congeners usually are in the same column of the periodic table.

control rod   a rod used in nuclear reactors to regulate the rate of the fission process by absorbing neutrons.

convection   the mixing of fluid or gas by the effect of warm material rising and cool material falling.

coordination compound   a compound that contains a central atom or ion and a group of ions or molecules surrounding it.

corrosion   degradation of a metal surface by a chemical or electro-chemical reaction.

covalent bond   a chemical bond formed by sharing valence electrons between two atoms (in contrast to an ionic bond, in which one or more valence electrons are transferred from one atom to another atom).

critical-point temperature   the temperature of a pure substance at which the liquid and gaseous phases become indistinguishable.

crucible   a dish used to heat a substance to high temperature; usually made of porcelain or metal.

cryogenic   refers to the production of very low temperatures and the study of the properties of materials at low temperatures.

crystallization   the process of forming crystals from liquids or solids.

cyanide   a compound or ion containing the CN^- group; such substances are extremely poisonous.

cyanide leaching   the use of a cyanide solution to dissolve gold from the surrounding rocks; the gold is subsequently recovered.

cyclotron   an apparatus that accelerates charged particles to very high energies as the particles travel in circular paths.

daughter isotope   an isotope produced by the radioactive decay of another (parent) isotope.

desalination   the removal of salt from seawater.

deuterium   an atom of heavy hydrogen that contains 1 proton, 1 neutron, and 1 electron.

deuteron   a nucleus of heavy hydrogen that contains 1 proton and 1 neutron.

dielectric constant   the capacity of a material to store electrical energy.

diffusion coefficient   a factor of proportionality that determines how fast a substance will diffuse across a barrier; differs for various substances.

disproportionation   an oxidation-reduction reaction in which some atoms of an element are oxidized while the remaining atoms of the same element are reduced.

doped   a material that has had an impurity introduced, usually for the purpose of improving conductivity in electronic circuits.

ductility   the ability of certain metals to be able to be drawn into thin wires without breaking.

effervescence   bubbling; the visible formation of a gaseous product during a chemical reaction taking place in solution.

ejecta   material ejected in supernova explosions.

eka-aluminum   Mendeleev's term for the element (now called gallium) he predicted would have properties similar to those of aluminum.

eka-boron   Mendeleev's term for the element (now called scandium) he predicted would have properties similar to those of boron.

eka-silicon   Mendeleev's term for the element (now called germanium) he predicted would have properties similar to those of silicon.

electrical conductivity   the ability of a substance, such as a metal, or a solution to conduct an electrical current.

electrolysis   the production of a chemical reaction achieved by passing an electrical current through an ionic solution.

electrolyte   a substance that, when dissolved in water, dissociates into ions sufficiently to conduct an electrical current.

electron   a subatomic particle found in all neutral atoms and negative ions; possesses the negative charges in atoms.

electronic configuration   a shorthand notation that indicates which atomic orbitals of an element's atoms are occupied by electrons and in what configuration.

electron-impact broadening   the widening of a spectral line owing to electrons impacting the atoms in a sample.

electron neutrino   the variety of neutrino associated with electron emission in beta decay.

electron pressure   the pressure produced in a dense star by closely packed electrons; the pressure prevents further collapse of the star.

electron volt   a unit of energy equivalent to the amount needed to move an electron across 1 volt of potential difference; abbreviation is eV.

electroplating   the process of coating one metal with another metal using electrolytic techniques.

electrostatic   the type of interaction that exists between electrically charged particles; electrostatic forces attract particles together if the particles have charges of opposite sign, while the forces cause particles that have charges of like sign to repel each other.

electrowinning   the technique of recovering a metal from its ore by using electrochemical means.

element   a pure chemical substance that contains only one kind of atom.

elution   the process of washing the compounds of a mixture through a chromatography column. (See *ion exchange chromatography*.)

energy   a measure of a system's ability to do work; expressed in units of joules.

enrichment   an increase in the percentage of one isotope of an element compared to the percentage that occurs naturally.

enzyme   one of a group of protein molecules that serve as catalysts of the chemical reactions taking place in living organisms.

erythronium   a variety of spring-flowering perennial.

espionage   the action of obtaining secret information; also called spying.

excimer   an excited state of a diatomic molecule that has no ground state.

fallout   the radioactive debris from a nuclear explosion or a nuclear accident.

family   (See *group*.)

fauna   animal life.

ferromagnetic   any metallic material that contains tiny magnetic domains that can be aligned by an external magnetic field.

filtration   the separation of substances in a mixture achieved by passing the mixture through a filter.

fission   (See *nuclear fission*.)

flavor   in particle physics, referring to distinct types of neutrinos associated with electrons, muons, or tauons.

flora   plant life.

fluorescence   the spontaneous emission of light from atoms or molecules when electrons make transitions from states of higher energy to states of lower energy.

fluorozirconate   an ion with the formula ZrF_6^{2-}.

fool's gold   the mineral iron pyrite (FeS_2); its shiny yellow color resembles gold.

fungicide   a substance that inhibits the growth of fungi.

fusion   may refer to a substance's phase change from solid to liquid or to nuclear fusion. (See *nuclear fusion*.)

galactic halo   the region at the outer edges of the Milky Way where very old metal-poor stars can be observed.

galvanometer   a device used to detect and measure electrical currents.

gamma decay   a mode of radioactive decay in which a very-high-energy photon of electromagnetic radiation—a gamma ray—is emitted; the daughter isotope has the same atomic number and mass number as the parent isotope, but lower energy.

gamma ray   a high-energy photon.

garnet   any of a group of silicate materials that may contain elements that include magnesium, calcium, manganese, iron, aluminum, chromium, or titanium; used as a gemstone or an abrasive.

Geiger counter   a device used to detect and measure the amount of radiation produced by radioactive decay.

germicide   a substance that kills germs.

glucose   a common simple sugar possessing the formula $C_6H_{12}O_6$; important source of energy in living organisms.

granite   an igneous rock that consists mostly of quartz and feldspar.

group   the elements that are located in the same column of the periodic table; also called a family, elements in the same column have similar chemical and physical properties.

gyroscope   a spinning device that helps maintain orientation.

half-life   the time required for half of the original nuclei in a sample to decay.

halogen   the elements in column VIIB of the periodic table; all of them share a common set of seven valence electrons in an nth energy level such that their outermost electronic configuration is ns^2np^5.

heat of fusion   the quantity of heat that a substance absorbs when undergoing the phase change from solid to liquid.

heat of vaporization   the quantity of heat that a substance absorbs when undergoing the phase change from liquid to vapor.

heavy water   water in which deuterium takes the place of hydrogen.

hemoglobin   a protein found in animals that carries oxygen through the bloodstream.

Hertzsprung-Russell diagram   used in astrophysics, a graph that plots luminosity versus surface temperature of a star.

heterogeneous catalyst   a catalyst that is in a different phase than the reaction being catalyzed; often a solid surface used to catalyze a reaction taking place in the gaseous or liquid phase.

hexavalent   describing an atom that can form six chemical bonds.

HgMn stars   a class of chemically peculiar stars that show a high abundance of mercury and manganese.

homogeneous catalyst   a catalyst that is in the same phase as the reaction being catalyzed.

hydride   a compound containing the H^- ion.

hydrosphere   the watery part of Earth; the oceans, lakes, rivers, icecaps, and glaciers.

hydrothermal vent   an opening in Earth's surface through which hot water issues.

igneous   a rock formed from the solidification of molten materials such as volcanic magma.

imaging   the process of producing pictures; in medicine, referring to pictures of bones or organs produced by methods such as X-ray photography.

incandescence   the emission of light that results from heating a substance to a high temperature.

inert   an element that has little or no tendency to form chemical bonds; the inert gases are also called *noble gases.*

infalling   materials falling into the atmosphere of a planet or star as a result of gravitational attraction.

ingot   a sample of metal or glass that has been cast into a shape suitable for further processing.

inner transition elements   the lanthanides and actinides.

insecticide   a substance intended to kill insects.

interstellar medium   gas and dust in interstellar space.

ion   an atom having a net electrical charge.

ion-exchange chromatography   the separation of different ions achieved by passing an ionic solution through a column containing a solid; different ions are absorbed by the solid and then released again at different rates, resulting in different times at which the ions exit the column.

ionic bond   a strong electrostatic attraction between a positive ion and a negative ion that holds the two ions together.

ionizing particle   a particle with properties (such as high energy or electrical charge) that allow it to ionize particles in any medium through which it passes.

isomeric state   one of many forms of an element with the same mass and atomic number but different radioactive states.

isotope   a form of an element characterized by a specific mass number; the different isotopes of an element have the same number of protons but different numbers of neutrons, hence different mass numbers.

joule the standard metric unit of energy; abbreviated J.

lanthanides the elements ranging from cerium (atomic number 58) to lutetium (number 71); they all have two outer electrons in the 6s subshell plus increasingly more electrons in the 4f subshell.

lanthanide series (See *lanthanides.*)

lapis lazuli a semiprecious stone whose main component is lazurite, with a unique color similar to cobalt blue, but usually of a deeper hue.

lattice the regular arrangement of atoms, molecules, or ions in a crystalline solid.

ligand an ion or molecule that donates a pair of electrons to a metal atom or ion.

lithosphere the solid part of Earth; depending on the context, may include Earth's crust, mantle, and core.

lodestone a black iron oxide (Fe_3O_4), also called *magnetite,* that possesses magnetic properties.

luminosity the brightness of a star defined as the total energy radiated per unit time.

magnetic domains in ferromagnetic materials, tiny regions that have locally the same magnetization.

magnetite (See *lodestone.*)

main group element an element in one of the first two columns or one of the right-hand six columns of the periodic table; distinguished from transition metals, which are located in the middle of the table, and from rare earths, which are located in the lower two rows shown apart from the rest of the table.

main sequence the area of the Hertzsprung-Russell diagram where most stars tend to be located during their early evolution.

malleability the ability of a substance such as a metal to change shape without breaking; metals that are malleable can be hammered into thin sheets.

mass   a measure of an object's resistance to acceleration; determined by the sum of the elementary particles comprising the object.

mass number   the sum of the number of protons and neutrons in the nucleus of an atom. (See *isotope*.)

metabolism   all of the chemical reactions that take place in living organisms.

metal   any of the elements characterized by being good conductors of electricity and heat in the solid state; approximately 75 percent of the elements are metals.

metal complex   (See *coordination compound*.)

metalloid (also called *semi-metal*)   any of the elements intermediate in properties between the metals and nonmetals; the elements in the periodic table located between metals and nonmetals.

metallurgy   the branch of engineering that deals with the production of metals from their ores, the manufacture of alloys, and the use of metals in engineering applications.

metamorphic   a rock that has been formed from a preexisting rock due to the application of heat, pressure, or chemical action.

metastable   an atomic state above the ground state that can be populated by the decay of an electron from a higher state. The state is relatively long-lived.

metavanadate   an ion with the formula VO_3^-.

mineralogist   a geologist who studies minerals.

mixture   a system that contains two or more different chemical substances.

molybdate   an ion with the formula MoO_4^{2-}.

mortality   death rate.

mutation   a random change in the genetic makeup of a cell.

nanometer   a metric unit of length equal to 1 billionth (10^{-9}) of a meter; abbreviated nm.

nanoparticle   any bit of matter whose diameter approaches less than 100 nanometers.

native metal   a metal found in its pure state instead of mixed with other elements in an ore.

negative image   the inverse of a normal picture; the exposure of film usually results in a negative image that is then printed as a normal, or positive, image.

neutrino   an elementary particle that has no charge and that travels at nearly the speed of light.

neutrino oscillation   the action by which one type of neutrino changes into a different type of neutrino.

neutron   the electrically neutral particle found in the nuclei of atoms.

neutron capture   the process in which a neutron collides with an atomic nucleus and is captured by that nucleus.

neutron degeneracy pressure   the pressure exerted by a population of neutrons that cannot be further compressed, according to the Pauli exclusion principle.

neutron star   a star that has reached the end of its evolutionary life cycle; under the force of gravity, all of the star's electrons are compressed into the nuclei, creating an extremely dense form of matter.

Ni-Cad battery   a rechargeable battery made of nickel and cadmium.

noble gas   any of the elements located in the last column of the periodic table—usually labeled column VIII or 18, or possibly column 0.

noble metal   a metal that is chemically resistant toward acids and bases; typically includes copper, silver, gold, mercury, and the platinum group of metals.

nonmetal   the elements on the far right-hand side of the periodic table that are characterized by little or no electrical or thermal conductivity, a dull appearance, and brittleness.

nuclear fission   the process in which certain isotopes of relatively heavy atoms such as uranium or plutonium spontaneously break apart into fragments; accompanied by the release of large amounts of energy.

nuclear fusion   the process in which nuclei lighter than iron can combine to form heavier nuclei; accompanied by the release of large amounts of energy.

nuclear medicine   the branch of medicine that uses radioactive isotopes for diagnosis or treatment of disease.

nucleon   a particle found in the nucleus of atoms; a proton or a neutron.

nucleosynthesis   the process by which atomic nuclei are synthesized.

nucleus   the small, central core of an atom.

nuclide   an atomic nucleus characterized by its numbers of protons and neutrons.

oxidation   an increase in an atom's oxidation state; accomplished by a loss of electrons or an increase in the number of chemical bonds to atoms of other elements. (See *oxidation state*.)

oxidation-reduction reaction   a chemical reaction in which one element is oxidized and another element is reduced.

oxidation state   a description of the number of atoms of other elements to which an atom is bonded. A neutral atom or neutral group of atoms of a single element are defined to be in the zero oxidation state. Otherwise, in compounds, an atom is defined as being in a positive or negative oxidation state, depending upon whether the atom is bonded to elements that, respectively, are more or less electronegative than that atom is.

oxidizing agent   a chemical reagent that causes an element in another reagent to be oxidized.

oxyanion   a negative ion that contains one or more oxygen atoms plus one or more atoms of at least one other element.

panchromism   exhibiting a wide range of colors.

parent isotope   an atom that undergoes radioactive decay into a daughter isotope.

passive   relatively chemically unreactive.

Pauli exclusion principle   initially put forth by Wolfgang Pauli in 1925, this principle states that two electrons having the same quantum numbers cannot occupy the same state in an atom.

pentahalide   a compound or ion that has five halogen atoms.

period   any of the rows of the periodic table; rows are referred to as periods because of the periodic, or repetitive, trends in the properties of the elements.

periodic table   an arrangement of the chemical elements into rows and columns such that the elements are in order of increasing atomic number, and elements located in the same column have similar chemical and physical properties.

pervanadyl   an ion with the formula VO_2^+.

pH   a measure of the acidity of an aqueous solution; low pH is strongly acid and high pH is strongly basic.

phase   referring to whether a substance is in the solid, liquid, or vapor state; also may refer to the different components of a mixture.

philosopher's stone   a substance believed in medieval times to have the power to convert base metals into gold.

phosphor   a substance that glows after being exposed to light.

photodisintegration   the breakup of nuclear material caused by collisions with high-energy photons.

photon   the name for the particle nature of light.

photovoltaic cell   a material used to produce electricity directly from sunlight.

physical change   any transformation that results in changes in a substance's physical state, such as color, temperature, dimensions, or other physical properties; the chemical identity of the substance remains unchanged in the process.

physical state   the condition of a substance being either a solid, liquid, or gas.

pig iron   iron that has been smelted in a coke furnace, resulting in a high carbon content.

platinum group   the elements platinum, palladium, ruthenium, rhodium, osmium, and iridium.

polyatomic   a molecule that contains two or more atoms or ions.

polymer   a substance composed of large molecules that consist of repeating units.

polymerization   a chemical reaction in which molecules join together to form a polymer.

post-transition metal   a naturally occurring metal located in the p-block of the periodic table: aluminum, gallium, indium, tin, thallium, lead, and bismuth.

potash   a common name for potassium carbonate (K_2CO_3).

precipitants   grains of metals that form in alloys.

primordial   anything associated with the beginning of time, either on Earth or in the universe.

product   the compounds that are formed as the result of a chemical reaction.

prosthetic   an artificial limb or orthopedic brace.

protein   a member of the group of large organic compounds that are found in living organisms and that contain *amino acids* as their basic structural unit.

Proterozoic   the period from approximately 2.5 billion to 570 million years before the present.

proton   the positively charged subatomic particle found in the nuclei of atoms.

qualitative analysis   the process of identifying what substances are present in a mixture.

quantitative analysis   the process of determining how much of a substance is present in a mixture.

quantum   a unit of discrete energy on the scale of single atoms, molecules, or photons of light.

quicksilver   an older name for the element mercury.

radioactive decay   the spontaneous disintegration of an atomic nucleus accompanied by the emission of a subatomic particle or gamma ray.

radiology   the use of ionizing radiation, especially X-rays, in medical diagnosis.

radionuclide   a radioactive nucleus.

rare earth element   the metallic elements found in the two bottom rows of the periodic table; the chemistry of their ions is determined by electronic configurations with partially filled *f* subshells. (See *lanthanides* and *actinides*.)

reactant   the chemical species present at the beginning of a chemical reaction that rearrange atoms to form new species.

red giant   the explosive evolutionary stage of a star that has fused all the hydrogen at its core into helium, which collapses under its own weight while hydrogen burning continues in the outer shell. The energy of the collapse generates radiation pressure that makes the outer shell expand explosively while the core continues to contract. The expanding gas in the outer shell filters out all but the star's red wavelengths.

reducing agent   a chemical reagent that causes an element in another reagent to be reduced to a lower oxidation state.

reduction   a decrease in an atom's oxidation state; accomplished by a gain of electrons or a decrease in the number of chemical bonds to atoms of other elements. (See *oxidation state*.)

roasting   the process of heating ores in a furnace to purify them.

rocky planet   one of the planets closest to the Sun—Mercury, Venus, Earth, and Mars.

r-process   the rapid capture by iron nuclei of a succession of neutrons, occurring during supernova explosions.

sacrificial anode   a substance that undergoes corrosion (thereby "sacrificing" itself) to protect an underlying metal; since corrosion is an oxidation process, the material that spontaneously undergoes corrosion is an electrochemical anode.

sandwich compound   a chemical compound in which a metal atom is "sandwiched" between two *aromatic hydrocarbons*.

scavenger   an animal that feeds on dead organic matter.

semiconductor   a material that relies on the jumping of electrons between accessible energy levels in order for current to flow.

semimetal   (See *metalloid*.)

semi-noble   an element that tends to resist undergoing chemical reactions.

sepsis   a bloodstream infection.

sesquioxide   a compound with the formula X_2O_3, where X is a metal ion with a "+3" charge.

shear tendency   the ability of a body to become deformed when subjected to a force parallel to one side of the body.

shell   all of the orbitals that have the same value of the principal energy level, notated as *n*.

silica   the compound SiO_2 as it is found in quartz, sand, and similar minerals.

smelter   the industrial location where pure metal is extracted from ore by various, usually temperature-dependent, separation processes; also refers to a person who works at a smelter.

solute   a substance present in lesser amount in a solution. A solution can have one or several solutes; for example, in seawater, each dissolved salt or gas is a solute.

solvent   the substance present in greatest amount in a solution; in *aqueous* solutions, water is the solvent.

specific heat   the heat per unit mass needed to raise the temperature of a substance by one degree.

spectral type   a designation given to a star indicating its brightness and color.

sphygmomanometer   a mercury-containing device for measuring blood pressure.

s-process   in massive stars, the so-called slow process, in which nuclei with masses greater than or equal to that of iron absorb neutrons to form heavier elements.

standard candle   an international measure of the intensity of light.

stardust   the small component of interstellar dust that thermally condenses from hot stellar vapor as it cools by expansion.

subatomic particle   particles that are smaller than atoms.

sublimation   the change of physical state in which a substance goes directly from the solid to the gas without passing through a liquid state.

subshell   all of the orbitals of a principal shell that lie at the same energy level.

superconducting   refers to superconducting materials, where current flow experiences virtually no resistance as it travels through the material, so no power is lost as heat.

supernova   a colossal explosive event ending the evolution of a high-mass star and ejecting its matter into interstellar space.

supersonic   traveling at a speed exceeding the speed of sound.

tailings   debris or waste material from mining operations.

tensile strength   the resistance of a material to forces that would tend to pull it apart.

teratogenic   a substance that causes birth defects.

tetravalent   describing an atom that can form four chemical bonds.

thermal conductivity   a measure of the ability of a substance to conduct heat.

thermal stability   a measure of the ability of a substance to resist the effects of a change in ambient temperature.

thio   a prefix used when an oxygen atom in an ion or compound has been replaced by a sulfur atom.

titanate   an ion in the form of TiO_3^{2-} or TiO_4^{2-}.

titration   a method of quantitative analysis in which a volume of one reagent, the concentration of which is known, is added to a known volume of another reagent, the concentration of which is unknown; when the two reagents have completely reacted, the unknown concentration can be calculated.

tracer   in medicine, a radioactive substance used to image problems such as tumors.

transactinide   an element that comes after lawrencium (element 103) in the periodic table.

transistor   an electronic circuit device used to modify signals.

transition metal   any of the metallic elements found in the 10 middle columns of the periodic table to the right of the alkaline earth metals; the chemistry of their ions is largely determined by electronic configurations with partially filled d subshells.

transmutation   the conversion by way of a nuclear reaction of one element into another element; in transmutation, the atomic number of the element must change.

transuranium element   any element in the periodic table with an atomic number greater than 92.

tritium   an isotope of hydrogen having two neutrons and one proton in the nucleus.

tungstate   an ion with the formula WO_4^{2-}.

type   a peculiar star.

type 304   the most versatile and widely used stainless steel, containing 18 percent chromium and 8 percent nickel.

ultraviolet   "beyond the violet"; light with wavelengths just shorter than those of visible light.

univalent   describing an atom that can form one chemical bond.

vaporization   the phase change in which a substance goes from the liquid to the vapor states.

weak s-process   a weak, slow process of neutron absorption in massive stars that produces some heavy elements.

white dwarf   a massive star in which fusion no longer can occur, it is supported by *electron pressure*. In two-star systems, a white dwarf accumulates matter from its companion star by gravitational attraction, leading to a Type I supernova explosion.

wrought iron   a form of iron that is not quite pure, in which melting and hammering is the method for forming objects.

X-ray   very-short-wavelength, high-frequency electromagnetic radiation; falls between ultraviolet light and gamma rays in frequency.

X-ray analysis   the use of X-rays to excite and identify elements.

X-ray fluorescence   a method for determining the content of various materials in which X-rays provide the needed analytical spectrum.

zircon   a mineral composed of zirconium silicate.

zirconate   an ion with the formula ZrO_3^{2-} or ZrO_4^{2-}.

Further Resources

THE SCANDIUM GROUP
Books and Articles

Macrakis, Kristie. *Seduced by Secrets.* Cambridge: Cambridge University Press, 2008. An important insider account of East German intelligence, including its use of radioactive tracers like scandium 46.

THE TITANIUM AND VANADIUM GROUPS
Books and Articles

Hurless, Brian E., and F. H. Froes. "Lowering the Cost of Titanium." *AMPTIAC Quarterly* 6, no. 2 (2002). This article from the U.S. Department of Defense Information Analysis Center describes the economics and market potential of titanium; extraction, melting, and casting costs; and applications and outlook for the future of the titanium industry.

Service, Robert. "Is Silicon's Reign Near Its End?" *Science* 323 (February 2009): 1001. This article explains hafnium's new role in the semiconductor industry.

Voorhees, Peter W. "Alloys: Scandium Overtakes Zirconium." *Nature Materials* 5 (June 2006): 435–436. This article explains the beneficial effects that can be achieved by the addition of zirconium to aluminum-scandium alloys.

Internet

United States Geological Survey. "Titanium Mineral Concentrates." Available online. URL: minerals.usgs.gov/minerals/pubs/commodity/titanium/mcs-2009-timin.pdf. Accessed December 4, 2009. This is the latest USGS information on titanium mineral concentrates as a world commodity, including production and use, prices, events, trends, and issues.

———. "Vanadium." Available online. URL: minerals.usgs.gov/minerals/pubs/commodity/vanadium/mcs-2009-vanad.pdf. Accessed

December 4, 2009. This is the latest USGS information on vanadium as a world commodity, including production and use, prices, events, trends, and issues.

THE CHROMIUM AND MANGANESE GROUPS
Books and Articles

Brennan, George. "Army to Scrap Tungsten Bullets." *Cape Cod Times,* 7 September 2009. This article discusses the discontinued use and sale of tungsten by the Department of Defense.

Hambling, David. "Study: 'Green' Training Ammo Carries Cancer Risk." *Wired,* 20 April 2009. This article discusses EPA plans for a closer look at tungsten's carcinogenic properties.

Internet

Environmental Protection Agency. "Consumer Fact Sheet on Chromium." Available online. URL: www.epa.gov/safewater/contaminants/dw_contamfs/chromium.html. Accessed December 4, 2009. Contains information on health effects of chromium and current regulation.

———. "Emerging Contaminant—Tungsten." Available online. URL: www.epa.gov/tio/download/contaminantfocus/epa542f07005.pdf. Accessed December 4, 2009. This fact sheet details current knowledge about tungsten in the environment and its potential health effects.

Lenntech Company Web Site. "Technetium—Tc." Available online. URL: www.lenntech.com/periodic/elements/tc.htm. Accessed December 4, 2009. This Web site describes the chemical properties and environmental and health effects of technetium.

Louisiana State University Chemistry Department. "Three-D Crystal Lattices." Available online. URL: www.chem.lsu.edu/htdocs/people/sfwatkins/ch4570/lattices/lattice.html. Accessed December 2, 2009. Descriptions and diagrams of different crystal lattice structures.

National Institutes of Health. "Hexavalent Chromium in Drinking Water Causes Cancer in Lab Animals." Available online. URL: www.nih.gov/news/pr/may2007/niehs-16.htm. Accessed December

4, 2009. This news release describes scientific findings about the effects of chromium-VI in drinking water.

Scientific American online. "California Acts to Control Chromium in Drinking Water." Available online. URL: www.scientificamerican. com/article.cfm?id=chromium-california-drinking-water. Accessed December 4, 2009. This article describes California's attempts to limit chromium levels in state water supplies.

United States Geological Survey. "Chromium." Available online. URL: minerals.usgs.gov/minerals/pubs/commodity/chromium/mcs-2009-chrom.pdf. Accessed November 29, 2009. This is the latest USGS information on chromium, from the U.S. Department of the Interior, as a world commodity, including production and use, prices, events, trends, and issues.

———. "Manganese." Available online. URL: minerals.usgs.gov/minerals/ pubs/commodity/manganese/mcs-2009-manga.pdf. Accessed November 29, 2009. This is the latest USGS information on manganese, from the U.S. Department of the Interior, as a world commodity, including production and use, prices, events, trends, and issues.

THE IRON, COBALT, AND NICKEL GROUPS
Books and Articles

Ball, Philip. *Universe of Stone.* New York: HarperCollins, 2008. This book describes the history of the cathedral at Chartres in exquisite detail. Important analysis of the cobalt blue windows is included.

Rehren, Thilo, and Edgar B. Pusch. "Late Bronze Age Glass Production at Qantir-Piramesses, Egypt." *Science* 308, no. 5,729 (17 June 2005): 1756–1758. This paper provides evidence for the production of glass from its raw materials in the eastern Nile Delta during the Late Bronze Age.

Internet

CBS news. "Cold Fusion Is Hot Again." April 19, 2009. Available online. www.cbsnews.com/stories/2009/04/17/60minutes/main4952167.shtml. Accessed November 25, 2009. This interview by *60 Minutes* correspondent Scott Pelley with researcher Michael

McKubre gives one scientist's opinion on what is going on in experiments with palladium and deuterium.

Hecht, Jeff. "Is Cold Fusion Heating Up?" *MIT Technology Review,* 23 April 2004. Available online. URL: www.technologyreview.com/energy/13559/page1/. Accessed November 25, 2009. This article gives an update on what scientists are learning about excess heat observed in experiments with palladium and deuterium.

United States Geological Survey. "Cobalt." Available online. URL: minerals.usgs.gov/minerals/pubs/commodity/cobalt/mcs-2009-cobal.pdf. Accessed November 29, 2009. This is the latest USGS information on cobalt as a world commodity, including production and use, prices, events, trends, and issues.

———. "Iron and Steel." Available online. URL: minerals.usgs.gov/minerals/pubs/commodity/iron_&_steel/mcs-2009-feste.pdf. Accessed November 29, 2009. This is the latest USGS information on iron and steel as world commodities, including production and use, prices, events, trends, and issues.

———. "Nickel." Available online. URL: minerals.usgs.gov/minerals/pubs/commodity/nickel/mcs-2009-nicke.pdf. Accessed November 29, 2009. This is the latest USGS information on nickel as a world commodity, including production and use, prices, events, trends, and issues.

Wren, Kathleen. "How Egypt Turned Dust into Treasures of Glass." Available online. URL: www.msnbc.msn.com/id/8221331/ns/technology_and_science-science. Accessed November 24, 2009. This article discusses archaeologists' discovery of a glass factory dating back to 1250 B.C.E.

THE COPPER GROUP

Books and Articles

Bernstein, Peter L. *The Power of Gold: The History of an Obsession.* New York: John Wiley & Sons, 2000. This book delves into the mythology and history of the quest for gold and its many uses, with a focus on how it inflames the greed and emotions of humans.

Gugliotta, Guy. "A Gold Mine's Toxic Bullet." *Washington Post,* 15 February 2000: A1. This article reports details about a cyanide spill from a gold mine in Romania that poisoned a river.

Harl, Kenneth W. *Coinage in the Roman Economy, 300 B.C. to A.D. 700.* Baltimore, Md.: The Johns Hopkins University Press, 1996. This book discusses the importance of various types of coins in ancient Roman society.

Lange, David W., and Mary Jo Mead. *History of the United States Mint and Its Coinage.* Atlanta, Ga.: Whitman Publishing, 2005. From the precolonial era to the present day, the history of the U.S. Mint is presented in a highly readable format.

Lewis, Nathan, and Addison Wiggin. *Gold: The Once and Future Money.* Hoboken, N.J.: John Wiley & Sons, 2007. In this book, the authors make the case for a return to the gold standard.

Schaps, David. *The Invention of Coinage and the Monetization of Ancient Greece.* Ann Arbor, Mich.: University of Michigan Press, 2007. This award-winning book describes how the invention of coinage as money was a Greek concept.

Internet

Johnson, C. A., D. J. Grimes, and R. O. Rye. "Accounting for Cyanide and Its Degradation Products at Three Nevada Gold Mines; Constraints from Stable C- and N-Isotopes." Available online. URL: pubs.er.usgs.gov/usgspubs/ofr/ofr98753. Accessed November 29, 2009. This U.S. Geological Survey Report (98-753) is a scientific look at the fate of cyanide in mine process waters.

Perlez, Jane, and Raymond Bonner. "Below a Mountain of Wealth, a River of Waste." *New York Times,* 17 December 2005. Available online. URL: www.nytimes.com/2005/12/27/international/asia/27gold.html. Accessed November 28, 2009. This article exposes extreme environmental damage to New Guinea's environment, resulting from political and industrial greed.

United States Geological Survey. "Copper." Available online. URL: minerals.usgs.gov/minerals/pubs/commodity/copper/mcs-2009-coppe.pdf. Accessed November 29, 2009. This is the latest USGS information on copper as a world commodity, from the U.S. Department of the Interior, including production and use, prices, events, trends, and issues.

———. "Gold." Available online. URL: minerals.usgs.gov/minerals/ pubs/commodity/gold/mcs-2009-gold.pdf. Accessed November 29, 2009. This is the latest USGS information on gold as a world commodity, from the U.S. Department of the Interior, including production and use, prices, events, trends, and issues.

———. "Silver." Available online. URL: minerals.usgs.gov/minerals/ pubs/commodity/silver/mcs-2009-silve.pdf. Accessed November 29, 2009. This is the latest USGS information on silver as a world commodity, from the U.S. Department of the Interior, including production and use, prices, events, trends, and issues.

THE ZINC GROUP
Books and Articles

Grissom, Carol A. *Zinc Sculpture in America: 1850–1950.* Cranbury, N.J.: Associated University Presses, 2009. This scholarly work explains the popularity of zinc as a sculpture material in the late 19th and early 20th centuries in the United States. History and technology, fabrication techniques, crafters and companies are discussed, and a catalogue of sculptures is included.

INTERNET

UNEP Global Mercury Partnership Newsletter, No. 1, October 2009. Available online. URL: www.chem.unep.ch/mercury/partnerships/ Hg_partnership_newsletter_Oct2009.pdf. Accessed December 1, 2009. United Nations Environment Programme news regarding plans for worldwide reduction of mercury releases.

World Health Organization Web Site. "Mercury in Health Care." August 2005. Available online. URL: www.who.int/water_sanitation_ health/medicalwaste/mercurypolpaper.pdf. A World Health Organization policy paper on the sources, effects, and alternatives to mercury use.

General Resources

The following sources discuss general information on the periodic table of the elements.

Books and Articles

Ball, Philip. *The Elements: A Very Short Introduction*. New York: Oxford University Press, 2004. This book contains useful information about the elements in general.

Chemical and Engineering News 86, no. 27 (2 July 2008). A production index published annually showing the quantities of various chemicals that are manufactured in the United States and other countries.

Considine, Douglas M., ed. *Van Nostrand's Encyclopedia of Chemistry*, 5th ed. New York: John Wiley and Sons, 2005. In addition to its coverage of traditional topics in chemistry, the encyclopedia has articles on nanotechnology, fuel cell technology, green chemistry, forensic chemistry, materials chemistry, and other areas of chemistry important to science and technology.

Cotton, F. Albert, Geoffrey Wilkinson, and Paul L. Gaus. *Basic Inorganic Chemistry*, 3rd ed. New York: Wiley and Sons, 1995. Written for a beginning course in inorganic chemistry, this book presents information about individual elements.

Cox, P. A. *The Elements on Earth: Inorganic Chemistry in the Environment*. New York: Oxford University Press, 1995. There are two parts to this book. The first part describes Earth and its geology and how elements and compounds are found in the environment. Also, it describes how elements are extracted from the environment. The second part describes the sources and properties of the individual elements.

Daintith, John, ed. *The Facts On File Dictionary of Chemistry*, 4th ed. New York: Facts On File, 2005. Definitions of many of the technical terms used by chemists.

Downs, A. J., ed. *Chemistry of Aluminium, Gallium, Indium and Thallium.* New York: Springer, 1993. A detailed, wide-ranging, authoritative and up-to-date review of the chemistry of aluminum, gallium, indium and thallium. Coverage is of the chemistry and commercial aspects of the elements themselves; emphasis is on the design and synthesis of materials, their properties, and applications.

Emsley, John. *Nature's Building Blocks: An A–Z Guide to the Elements.* New York: Oxford University Press, 2001. Proceeding through the periodic table in alphabetical order of the elements, Emsley describes each element's important properties, biological and medical roles, and importance in history and the economy.

———. *The Elements.* New York: Oxford University Press, 1989. In this book, Emsley provides a quick reference guide to the chemical, physical, nuclear, and electron shell properties of each of the elements.

Foundations of Chemistry 12, no. 1 (April 10, 2010). This special issue of the journal focuses on the periodic table, featuring some obscure history, possible new arrangement, and the role of chemical triads.

Greenberg, Arthur. *Chemistry: Decade by Decade.* New York: Facts On File, 2007. An excellent book that highlights by decade the important events that occurred in chemistry during the 20th century.

Greenwood, N. N., and A. Earnshaw. *Chemistry of the Elements.* Oxford, U.K.: Pergamon Press, 1984. This book is a comprehensive treatment of the chemistry of the elements.

Hall, Nina, ed. *The New Chemistry.* Cambridge: Cambridge University Press, 2000. Contains chapters devoted to the properties of metals and electrochemical energy conversion.

Hampel, Clifford A., ed. *The Encyclopedia of the Chemical Elements.* New York: Reinhold Book, 1968. In addition to articles about individual elements, this book also has articles about general topics in chemistry. Numerous authors contributed to this book, all of whom were experts in their respective fields.

Heiserman, David L. *Exploring Chemical Elements and Their Compounds.* Blue Ridge Summit, Penn.: Tab Books, 1992. This book is described by its author as "a guided tour of the periodic table for ages 12 and up," and is written at a level that is very readable for precollege students.

Henderson, William. *Main Group Chemistry.* Cambridge, U.K.: The Royal Society of Chemistry, 2002. This book is a summary of inorganic chemistry in which the elements are grouped by families.

Jolly, William L. *The Chemistry of the Non-Metals.* Englewood Cliffs, N.J.: Prentice-Hall, 1966. This book is an introduction to the chemistry of the nonmetals, including the elements covered in this book.

King, R. Bruce. *Inorganic Chemistry of Main Group Elements.* New York: Wiley-VCH, 1995. This book describes the chemistry of the elements in the *s* and *p* blocks.

Krebs, Robert E. *The History and Use of Our Earth's Chemical Elements: A Reference Guide,* 2nd ed. Westport, Conn.: Greenwood Press, 2006. Following brief introductions to the history of chemistry and atomic structure, Krebs proceeds to discuss the chemical and physical properties of the elements group (column) by group. In addition, he describes the history of each element and current uses.

Lide, David R., ed. *CRC Handbook of Chemistry and Physics,* 89th ed. Boca Raton, Fla.: CRC Press, 2008. The *CRC Handbook* has been the most authoritative, up-to-date source of scientific data for almost nine decades.

Mendeleev, Dmitri Ivanovich. *Mendeleev on the Periodic Law: Selected Writings, 1869–1905.* Mineola, N.Y.: Dover, 2005. This English translation of 13 of Mendeleev's historic articles is the first easily accessible source of his major writings.

Minkle, J. R. "Element 118 Discovered Again—For the First Time." *Scientific American,* 17 October 2006. This article describes how scientists in California and Russia fabricated element 118.

Norman, Nicolas C. *Periodicity and the p-Block Elements.* New York: Oxford University Press, 1994. This book describes group properties of post-transition metals, metalloids, and nonmetals.

Parker, Sybil P., ed. *McGraw-Hill Encyclopedia of Chemistry,* 2nd ed. New York: McGraw Hill, 1993. This book presents a comprehensive treatment of the chemical elements and related topics in chemistry, including expert-authored coverage of analytical chemistry, biochemistry, inorganic chemistry, physical chemistry, and polymer chemistry.

Rouvray, Dennis H., and R. Bruce King, ed. *The Periodic Table: Into the 21st Century.* Baldock, Hertfordshire, U.K.: Research Studies Press Ltd., 2004. A presentation of what is happening currently in the world of chemistry.

Stwertka, Albert. *A Guide to the Elements,* 2nd ed. New York: Oxford University Press, 2002. This book explains some of the basic concepts of chemistry and traces the history and development of the periodic table of the elements in clear, nontechnical language.

Van Nostrand's Encyclopedia of Chemistry, 5th ed., Glenn D. Considine, ed. Hoboken, N.J.: Wiley and Sons, 2005. This is a compendium of modern chemistry covering topics from green chemistry to nanotechnology.

Winter, Mark J., and John E. Andrew. *Foundations of Inorganic Chemistry.* New York: Oxford University Press, 2000. This book presents an elementary introduction to atomic structure, the periodic table, chemical bonding, oxidation and reduction, and the chemistry of the elements in the *s, p,* and *d* blocks; in addition, there is a separate chapter devoted just to the chemical and physical properties of hydrogen.

Internet Resources

About.com: Chemistry. Available online. URL: chemistry.about. com/od/chemistryfaqs/f/element.htm. Accessed December 4, 2009. Information about the periodic table, the elements, and chemistry in general from the New York Times Company.

American Chemical Society. Available online. URL: portal.acs.org/ portal/acs/corg/content. Accessed December 4, 2009. Many educational resources are available here.

Center for Science and Engineering Education, Lawrence Berkeley Laboratory, Berkeley, California. Available online. URL: www.lbl. gov/Education. Accessed December 4, 2009. Contains educational resources in biology, chemistry, physics, and astronomy.

Chemical Education Digital Library. Available online. URL: www.chemeddl.org/index.html. Accessed December 4, 2009. Digital content intended for chemical science education.

Chemical Elements. Available online. URL: www.chemistryexplained.com/elements. Accessed December 4, 2009. Information about each of the chemical elements.

Chemical Elements.com. Available online. URL: www.chemical elements.com. Accessed December 4, 2009. A private site that originated with a school science fair project.

Chemicool. Available online. URL: www.chemicool.com. Accessed December 4, 2009. Information about the periodic table and the chemical elements, created by David D. Hsu of the Massachusetts Institute of Technology.

Department of Chemistry, University of Nottingham, United Kingdom. Available online. URL: www.periodicvideos.com. Accessed December 4, 2009. Short videos on all of the elements can be viewed.

Journal of Chemical Education, Division of Chemical Education, American Chemical Society. Available online. URL: jchemed.chem.wisc.edu/HS/index.html. Accessed December 4, 2009. The Web site for the premier online journal in chemical education.

Lenntech. Available online. URL: www.lenntech.com/Periodic-chart.htm. Accessed December 4, 2009. Contains an interactive, printable version of the periodic table.

Los Alamos National Laboratory. Available online. URL: periodic.lanl.gov/default.htm. Accessed December 4, 2009. A resource on the periodic table for elementary, middle school, and high school students.

Mineral Information Institute. Available online. URL: www.mii.org. Accessed December 4, 2009. A large amount of information for teachers and students about rocks and minerals and the mining industry.

National Nuclear Data Center, Brookhaven National Laboratory. Available online. URL: www.nndc.bnl.gov/content/HistoryOfElements.html. Accessed December 4, 2009. A worldwide resource for nuclear data.

The Periodic Table of Comic Books, Department of Chemistry, University of Kentucky. Available online. URL: www.uky.edu/Projects/Chemcomics. Accessed December 8, 2009. A fun, interactive version of the periodic table.

The Royal Society of Chemistry. Available online. URL: http://www.rsc.org/chemsoc/. Accessed December 4, 2009. This site contains information about many aspects of the periodic table of the elements.

Schmidel & Wojcik: Web Weavers. Available online. URL: quizhub.com/quiz/f-elements.cfm. Accessed December 4, 2009. A K–12 interactive learning center that features educational quiz games for English language arts, mathematics, geography, history, earth science, biology, chemistry, and physics.

United States Geological Survey. Available online. URL: minerals.usgs.gov. Accessed December 4, 2009. The official Web site of the Mineral Resources Program.

Web Elements, The University of Sheffield, United Kingdom. Available online. URL: www.webelements.com/index.html. Accessed December 4, 2009. A vast amount of information about the chemical elements.

Wolfram Science. Available online. URL: demonstrations.wolfram.com/PropertiesOfChemicalElements. Accessed December 4, 2009. Information about the chemical elements from the Wolfram Demonstration Project.

Index

Note: *Italic* page numbers refer to illustrations.